HANDBOOK OF ELECTROSTATIC DISCHARGE CONTROLS (ESD)

Facilities Design and Manufacturing Procedures

Bernard S. Matisoff, P.E., CMfgE

VNR VAN NOSTRAND REINHOLD COMPANY
New York

Library of Congress Catalog Card Number: 85-638
ISBN: 0-442-26071-7

Manufactured in the United States of America

Published by Van Nostrand Reinhold Company Inc.
115 Fifth Avenue
New York, N.Y. 10003

Van Nostrand Reinhold Company Limited
Molly Millars Lane
Wokingham, Berkshire RG11 2PY, England

Van Nostrand Reinhold
480 Latrobe Street
Melbourne, Victoria 3000, Australia

Macmillan of Canada
Division of Gage Publishing Limited
164 Commander Boulevard
Agincourt, Ontario M1S 3C7, Canada

15 14 13 12 11 10 9 8 7 6 5 4 3 2 1

Library of Congress Cataloging in Publication Data

Matisoff, Bernard S.
Handbook of electrostatic discharge controls (ESD)
Includes index.
1. Static eliminators—Handbooks, manuals, etc.
2. Electrostatics—Handbooks, manuals, etc. I. Title.
TK7835.M35 1985 621.31′028′9 85-638
ISBN 0-442-26071-7

To Donald E. Brown, my friend, in appreciation for his support and confidence over these many years.

PREFACE

By damaging electronic components, *static electricity* could conceivably immobilize an aircraft or warship, not to mention a fire control or countermeasure system. It can also have the same effect on commercial systems such as computers and industrial controls. If the oxide gate of a semiconductor device is struck by a miniature lightning bolt from a static discharge, it may promptly be rendered useless, or worse, may fail prematurely in the field, even though it may have passed all of the in-house testing prior to shipment or use in an electronic assembly. Static electricity is a fact of life and is around us at all times. It cannot be eliminated, but it can be controlled.

The need for a comprehensive engineering handbook on the subject of electrostatic controls has been evident for the past several years. Although this handbook has a definite orientation toward the electronics manufacturing industry, the fundamental techniques described are equally applicable to other fields in which static damage is a known factor, such as the medical field, computer processing, electronic controls and instrumentation, and those areas where explosive materials are manufactured or used. Although the engineering level of the handbook is high, an attempt has been made to make it readily understandable by the practical technical personnel who will actually install and use these materials and methods. In short, the overall treatment is practical rather than classical so as better to serve practicing engineers in their day-to-day work. Rigorous mathematical derivations have been avoided.

If one were to read everything available on the subject of static electricity—papers, magazines, catalogs, and so forth—one might indeed find everything that is contained in this text. It is believed, however, that this is the first time that almost all of the practical information on the subject has been assembled in a way that permits the reader to apply it immediately.

Bernard S. Matisoff

ACKNOWLEDGMENT

I would like to thank the Bemis Company, Inc., of Minneapolis, Minnesota, the Bengal Company, Inc., of Sepulveda, California, and the 3M Static Controls Division of Saint Paul, Minnesota, for their contributions to the compilation of the data in this text. Since every salesman has his own convincing reasons why his product is the *only* one that will solve the static problem, without their assistance the selection of useful data from the vast pool of information on this subject would have been impossible.

Bernie Matisoff, CMfgE, P.E.

CONTENTS

1. INTRODUCTION

THE PROBLEM OF ESD

Electrostatic discharge, or ESD, is one of the most destructive yet misunderstood phenomena in the electronics industry today. While ESD cannot be eliminated, its destructive effects can be controlled.

The intent of this book is to provide the engineer with the means for generating and implementing a static control system with sufficient understanding of the applications and uses of the available products in the marketplace.

In most instances the design of a static-controlled facility and the generation of the manufacturing procedures is assigned to a manufacturing engineer. The problems that arise are due to the manufacturing engineer's lack of understanding of the electronic principles involved. The intent here is to provide these engineers with a sufficient understanding of the principles of static control and the methods employed in developing a facility in which high reliability electronics can be manufactured.

Today's challenge within the electronics industry is (1) to miniaturize electronic systems by producing smaller lines on closer spacing, sandwiching series of printed circuit boards together to create a more compact system, and (2) to employ flexible circuits for specific applications.

The first prerequisite in the manufacture of modern printed circuits and flexible circuits is a controlled atmosphere containing little or no contamination. The photo-imaging of printed circuit boards cannot be accomplished on a surface that has dirt on it. Contamination between prepregs and boards, or in the holes of multilayer boards, can and does result in shorts and opens in the circuitry. Yet most manufacturers today operate in the traditional plant environment where static electricity is prevalent. The cost of product contamination caused by airborne particulates which are very difficult to remove can be high. In addition to the contamination problems, the failure to acknowledge the sensitivity of modern IC devices contributes to a continuous increase in production line failures. In response to the electrostatic discharge problems, field service staffs are increased at considerable cost, while the basic problem exists back at the production line. Sensitive devices must be protected.

As device technology pushes forward to achieve higher speeds and greater functional density, device geometries are shrinking. Line widths and spaces are getting smaller, and oxide layers are getting thinner. Consequently, lower voltages and current will damage these sensitive parts.

Static electricity is a fact of nature; it is all around us at all times. Day after day, in electronic manufacturing plants all over the world, this invisible process causes the loss of millions of dollars, yet it is one of the least understood and least recognized effects haunting the industry today. The process might begin when workers who handle static-sensitive electronic components such as MOSFETs, JFETs, CMOSs, bipolar transistors, and many more report to work wearing a pair of crepe-soled shoes. Most factories have a nonconductive floor and no means of controlling the humidity. As the humidity drops below 20% the workers, just walking to their bench, will build up a static electric charge on their body. This, known as the triboelectric effect, is due to the workers' shoes repeatedly making and breaking contact with the floor.

The faster the workers walk, the higher the charge, and the stronger the field around their body. As they arrive at their workbench, which may have several static-sensitive devices on it, the static field around them reaches the bench before them and induces a current to flow through the devices, at which point the static-sensitive devices could be destroyed or damaged.

Once the worker sits down, his field is now close enough to cause the devices on the bench to become polarized, so that damage could again result without the devices having even been touched. Then, as the worker reaches out to pick up one of the devices, a spark jumps from his fingertip to the device and *zap,* there is damage for sure. This, known as electromagnetic pulse, could damage all the nearby devices in its field.

These are only some of the conditions encountered in electronics manufacturing. All of them must be understood and controlled before the large losses currently experienced can be eliminated. The cost of damage and rework on circuits due to zapping has been estimated at half a billion dollars per year. Experts state that an annual potential of

$5 billion could be saved through the elimination of static-related damage. Top engineers believe that up to 25% of all component failures are caused by ESD.

If the zap is discovered at the chip stage, it may cost only $1.50 to replace. If it's at the board stage (assembled chips), it could be from $15 to $30, while if it's already in a piece of machinery, costs in the thousands could result, or product failure that has led to the accidental death of pilots and technicians. The real danger is when the zap causes a latent defect instead of outright failure. The latter can be tested and caught in inspection, whereas the former causes problems in the field when the product fails prematurely.

Awareness and recognition of the potential damage is nine-tenths of the problem. You can't feel a 1,000-volt static discharge—or see it. But 1,000 volts can cause a very large number of electronic component failures. Electronic components are most vulnerable at the chip or integrated-circuit stage. However, when many ICs are assembled on a board or even within the final product, zaps can occur.

There are presently two methods of identifying the ESD sensitivity of electronic components. U.S. DOD-STD-1686 recognizes two classes of ESD-sensitive items:

Class I	100 to 1,000 volts sensitivity
Class II	1,001 to 4,000 volts sensitivity

Some private industry groups use four classes:

Class 0	0 to 170 volts
Class 1	171 to 1,000 volts
Class 2	1,001 to 4,000 volts
Class 3	4,001 to 15,000 volts

Considerable confusion and misunderstanding of static electricity are created by the salesmanship of the many companies manufacturing static control products. Each salesman has a convincing reason why his product is the only one that will do the job. Because of this, many customers have spent a fortune to eliminate or control static electricity, yet the problem still exists.

The engineer faced with designing and implementing a static control system must understand the problem and be aware of the available solutions. He must evaluate the system as a total, all the way from the handling and shipping of critical parts to his plant by the supplier, to the shipping of the final product. To meet the challenge, static control must become a way of life where control is the key. Not simply *static* control, but control and management of the static problem.

Effective static management is based on three fundamentals. First, it recognizes static as an environmental problem—one that extends beyond the boundaries of a work station and affects most functional areas in any facility. Components are as susceptible to ESD in receiving as in board assembly, and as susceptible on carts as on work benches.

Second, effective static management attacks the causes of static rather than its symptoms. The elimination of static-producing materials from the work area is more effective than permitting their presence, and then relying on grounding or ionization to eliminate the charge they generate.

Third, effective static management requires a comprehensive program recognizing that no single product or procedure can do the job alone or be 100% effective. Wrist straps, ionization, packaging materials, and other products or procedures are synergistic and complementary: their total effect is greater than the sum of each independently, since each one also backs up the others.

As already noted, one of the primary generators of static in any populated environment is the movement of personnel or equipment across a floor. The interaction between shoe or caster and floor can generate significant static voltages (see Table 1-1).

Static would be less of a problem if personnel were stationary—if they didn't take coffee or lunch breaks or move products from one area to another, and if supervisors could supervise from their offices. But movement exists and static is generated.

Table 1-1. Typical Electrostatic Voltages.

MEANS OF STATIC GENERATION	ELECTROSTATIC VOLTAGES	
	10-20% R.H.	65-90% R.H.
Walking across carpet	35,000	1,500
Walking over vinyl floor	12,000	250
Worker at bench	6,000	100
Mobile storage carts on vinyl floors	Up to 5,000 volts	

Personnel-generated static is commonly controlled with a wrist strap which is frequently considered the first line of defense. The wrist strap works as intended but doesn't completely control the problem, because it protects only at the work station and doesn't remove static charge from clothing. Personnel approaching a work station carry potentially damaging static charges until a wrist strap is attached.

Reliance upon wrist straps alone to control static in personnel and equipment addresses a symptom rather than the cause. It contributes little to environmental static control, and leaves ample opportunities for static problems.Table 1-2 shows a few of the major sources of static electricity.

Table 1-2. Typical Prime Charge Sources.

OBJECT OR PROCESS	MATERIAL OR ACTIVITY
Work surfaces	• Waxed, painted or varnished surfaces • Common vinyl or plastics
Floors	• Sealed concrete • Waxed, finished wood • Common vinyl tile or sheeting
Clothes	• Common clean room smocks • Common synthetic personnel garments • Nonconductive shoes • Virgin cotton[a]
Packaging and handling	• Common plastic: bags, wraps, envelopes • Common bubble pack, foam • Common plastic trays, plastic tote boxes, vials, parts bins
Assembly, cleaning, test, and repair areas	• Spray cleaners • Common plastic solder suckers • Solder irons with ungrounded tips • Solvent brushes (synthetic bristles) • Cleaning or drying by fluid or evaporation • Temperature chambers • Cryogenic sprays • Heat guns and blowers • Sand blasting • Electrostatic copiers

[a] Virgin cotton can be a static source at low relative humidities such as below 30%.

THE SYSTEM—IT WORKS

Although the manufacturers of ESD-protective materials and devices may disagree on the applications of conductive versus antistatic approaches, there is one point on which they all agree: the approach to static control must be thorough and systematic, it must include the entire manufacturing cycle, or it will not work. An effort must be made to minimize ESD in every aspect of the manufacturing process, from assembly and handling to storage and shipping. If any phase is omitted, one might just as well throw away all the special materials and devices, and in many cases the printed circuit boards with them.

Static-charge-induced failures are a well-documented fact—one of considerable proportions. So how can one protect parts and assemblies from ESD? Two basic solutions have evolved from the current studies.

1. If it moves, shield it.
2. If it doesn't move, ground it.

Both these solutions can be very effective. *Can* in this instance is based on a number of *ifs:* if shields are effective during their life cycle; if grounds are truly grounded; if every point in the manufacturing cycle is constantly controlled—from receipt of incoming parts and materials through the completion and shipping of the final product.

SUMMARY

The antistatic approach to ESD control is the best and most cost-effective approach in most applications for protection of sensitive devices and assemblies. If the device warrants greater precaution, the best protection is achieved by placing ESD-sensitive parts in an antistatic bag or container and then inserting the antistatic bag into an aluminum foil bag. This prevents damage from handling when it is removed from the aluminum foil bag, and protects the device or assembly from static fields, ESD, RFI, and EMI.

The term *caveat emptor* (buyer beware) has no greater significance than in the procurement of materials and equipment for ESD control. Most purchases do not come directly from reputable manufacturers, but from wholesalers or distributors who represent the materials as they are represented to them, and in many instances do not know if they are static-free. There are many pink poly and conductive materials and several blue materials available, and we can be certain there will be a multitude of copies and counterfeits as the importance of ESD control achieves greater recognition in the future.

STATIC CONTROL CHECKLIST

1. Receiving/Receiving Inspection
   - Bench ______
   - Table mat ______
   - Floor mat ______
   - Wrist strap ______
2. Storage
   - Bins ______
   - Racks ______
   - Shelves ______
   - IC Tubes ______
   - Cabinets ______
3. Kitting
   - Bench ______
   - Table mat ______
   - Floor mat ______
   - Wrist strap ______
   - Tote boxes ______
   - Bags ______
   - Bins ______
4. Assembly
   - Benches ______
   - Table mat ______
   - Floor mat ______
   - Wrist strap ______
   - Clothing ______
   - Ionizer ______
   - Bins ______
   - Tote boxes ______
   - Shoe straps ______
   - Stool covers ______
   - Bags ______

STATIC CONTROL CHECKLIST *(continued)*

5. Transportation
   - Carts __________
   - Tote boxes __________
   - Bags __________
   - IC Tubes __________
   - Wrist strap, grounded to cart __________
   - Ground strap, cart to floor __________

6. Work in process (WIP Stores)
   - Cabinets __________
   - Shelves __________
   - Bins __________
   - Tote boxes __________
   - Bags __________
   - Covers, antistat for documentation __________

7. Inspection
   - Bench __________
   - Table mat __________
   - Floor mat __________
   - Wrist strap __________
   - Ionizer __________
   - Clothing __________
   - Tote Boxes __________
   - Bins __________
   - Bags __________

8. Shipping
   - Bench __________
   - Table mat __________
   - Floor mat __________
   - Wrist strap __________
   - Ionizer __________
   - Antistat boxes __________
   - Antistat foam __________
   - Antistat wrapping __________

9. Testing
   - Same as Assembly __________

2. THE NATURE OF ELECTROSTATIC DISCHARGE (ESD) CONTROL

ESD DEFINED

Electrostatic discharge is defined in the U.S. military handbook DOD-HKBK-263 as a "transfer of electrostatic charges between bodies at different potentials caused by direct contact or induced by an electrostatic field." This transfer of electrostatic charge causes destruction to electronic components, which today are becoming more susceptible than ever before.

Modern manufacturing, quality assurance, and product reliability have been confronted with the additional—though not thoroughly understood—problems of electrostatic discharge. ESD is the primary source of damage and failures to semiconductor devices and can occur at any time during the manufacturing cycle.

STATIC ELECTRICITY

To properly select materials and equipment so as to prevent ESD damage, an understanding of the basic functional physics of static-free production is necessary. One must understand what occurs when a static charge is applied or induced upon a static-free component. The most important factor to remember is that when static electricity ceases to be at rest, it is no longer "static." Static charge in motion becomes an electrical charge-producing current and power, or heat.

Electrostatic charge on a body can be imparted to another body through induction and conduction. This is important, since it is static's ability to create charges on other objects by these mechanisms that makes it so dangerous. Electrostatic charges on ungrounded conductors will distribute evenly over the entire surface. A charged conductor which touches another uncharged, ungrounded conductor will transmit some of its charge. This charge transmission continues until the charge distribution reaches an equilibrium point. Thus the uncharged conductor has become charged, while the originally charged conductor loses some of its charge. On the other hand, charges on nonconductors will remain at the point where they were initially formed, even if the material is connected to a ground.

A common example of static is found in almost every home. After washing clothes, one puts them in a dryer. As the dryer removes all the moisture, the synthetic fibers—usually polyesters—develop a surface resistance so great that static electricity is generated. To prevent "static cling," a strip of paper containing an antistatic agent is placed in the dryer along with the clothes. This strip of paper lowers the surface resistivity of the fibers to prevent the generation of static electricity. However, since most antistatic agents are soap-based chemicals, adding a small amount of soap to the washer during the rinse cycle would also eliminate static electricity.

Static damage to IC components by operating personnel is becoming one of the most significant problems plaguing the electronics industry. Technological advances in manufacture such as ion implantation, ion beam milling, and electron beam direct writing all make possible devices with higher circuit densities, higher unit performance, and often higher static susceptibility. Some are more apt to be damaged by static charges than others, and the list of components considered highly ESD-sensitive is growing every year. Table 2-1, a summary of some exist-

Table 2-1. Susceptibility Ranges of Devices Exposed to Electrostatic Discharge.

DEVICE TYPE	RANGE OF ESD SUSCEPTIBILITY (VOLTS)
Bipolar transistors	380-7,000
CMOS	250-3,000
ECL	500-1,500
EPROM	100
Film resistors (thick, thin)	300-3,000
GaAsFET	100- 300
JFET	140-7,000
MOSFET	100-200
OP-AMP	190-2,500
SAW	150-500
Schottky diodes	300-2,500
Schottky TTL	1,000-2,500
SCR	680-1,000
VMOS	30-1,800

ing data from various industry sources, shows the approximate range of static susceptibility for several different component types.

Static electricity is a difficult subject at best to discuss. Its appearance in the industrial environment on a large scale is relatively new. Static electricity manifests its destructive characteristics in two distinctive ways. The first is *magnetic attraction,* which occurs when airborne particles of dirt become charged (negative or positive). This can happen for many reasons: movement of air through an air-conditioning system or a heating system, the presence of people, the presence of static-charged materials, and so on. Once charged, the particles are attracted to a surface with the opposite charge. Since the charges are both positive and negative, contamination is drawn to all types of surfaces within the environment. Once attracted, the same magnetic force holds the contaminants *fixed* to a surface. Brushing, wiping, blowing with air, and so forth will not remove the contamination, which must be neutralized. A surface that is neutralized will *not* attract contamination.

The second characteristic of static electricity is *ESD* or *electrostatic discharge.* The electrostatic buildup on people or materials—particularly nonconductive materials—can be significant. The average individual walking across a nonconductive floor can generate from 3,000 to 4,000 volts; a plant environment with a forced air-heating or an air-conditioning system will generate the same level of charge. Plastics used in many products will produce charges of from 5,000 to 10,000 volts.

The failure to acknowledge the sensitivity of IC devices results in a continuing increase in production-line failures. The problem of lost production has reached significant economic levels. Field and customer warranty claims are skyrocketing.

Too often, problems associated with static electricity are approached on a piecemeal basis. The first vendor through the door selling a cure-all is embraced. The recommended solution is implemented with no tangible results, which is costly. Prior to implementing a solution, the user must address and define the problem.

Static electricity is charge at rest on an object caused by a transfer of electrons. This charge can be either negative, if there is an excess of electrons on a body, or positive, if there is a deficiency of electrons on that body. Electron transfer occurs when two contacting bodies are rubbed together, separated, or flow past each other—a phenomenon known as the triboelectric effect. Electrostatic charges on ungrounded conductors will distribute evenly over the entire surface. Charges on nonconductors remain at the point where they were initially formed, even if the material is connected to a ground.

ESD—CAUSE AND EFFECT

The root cause of electrostatic discharge is friction. The heat of friction excites the electrons in the outer orbits of molecules, causing them to transfer from surface to surface. If we could eliminate friction, we would automatically eliminate static electricity. Since this is impossible, we must learn to control the reactions created by friction and thus control the generation of static electricity.

As two surfaces are rubbed together, the heat of friction causes the electrons in the outer orbit to transfer from surface to surface. Friction initiates the start of an ESD charge buildup. The surface that surrenders the electrons will suffer from a deficiency of electrons, and positive ions (a positive charge) result. The opposite surface that is in contact with the above-mentioned surface will accept these electrons; an excess of electrons, therefore, results in a negative charge.

Let us translate the above into practical terms for the use of conductive and nonconductive materials. These terms are confusing and too often misunderstood. If a nonconductive tote tray is slid over a grounded conductive surface, electrons can transfer. If the electrons transfer from the nonconductive tote box to the conductive surface, the nonconductive tote box will take on a positive charge. However, the conductive surface will not charge, being conductive and tied to earth ground. The excess of electrons will flow to ground and the conductive surface will not charge. However, the nonconductive tote box will maintain its charge, since it is not a conductor of electricity, hence the term "static electricity."

Significant amounts of static electricity can be generated in the electronics industry, owing to the use of static-generator-type processes and methods. The ideal solution to ESD control would be to eliminate all static-generating sources. This, of course, would mean the elimination of people and virtually all movement. In the real world in which we function, this is not possible.

Static electricity can be controlled in any operation. To effectively do so requires detailed analysis, planning, effective execution of the plan, and discipline. Employees must be educated in both the problems and the solutions to the generation and control of static electricity. Further, everyone must be made to realize that *static electricity cannot be eliminated,* it can only be controlled.

As technology pushes to achieve higher speeds and greater functional density, device geometries are shrinking, line widths and spaces are getting smaller, and oxide layers are getting thinner. Consequently, lower voltages and currents will damage these sensitive parts (Figure 2-1).

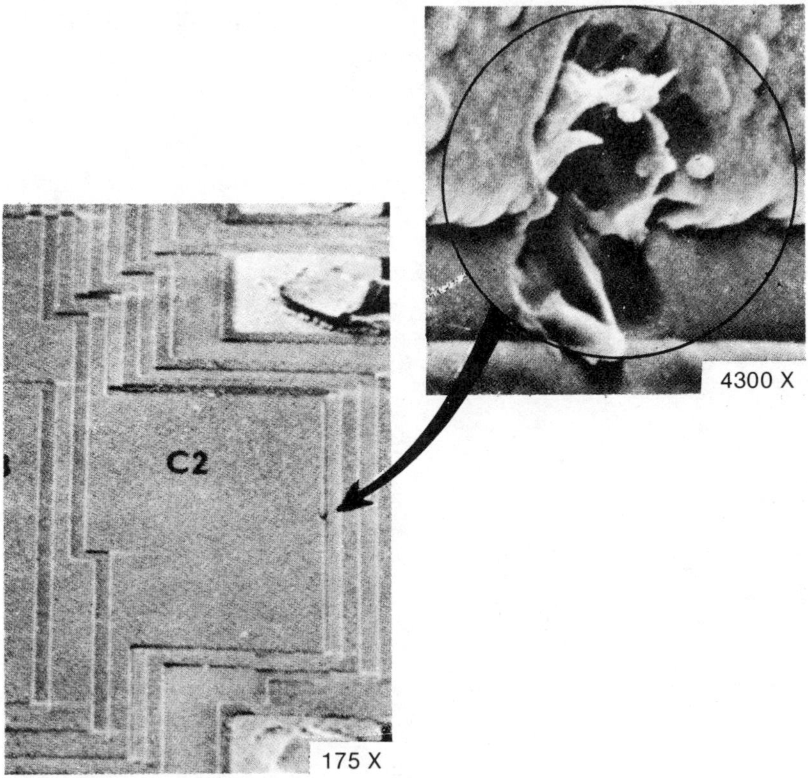

Figure 2-1. A highly magnified closeup showing damage caused by ESD.

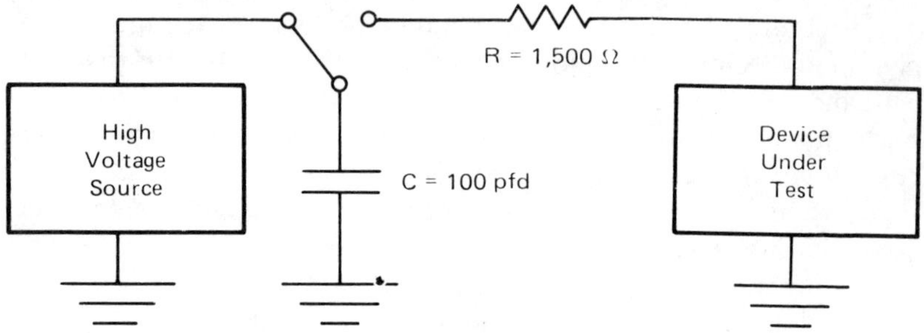

Figure 2-2. The human-body model.

The human-body model is the discharge simulation model most frequently used in testing ESD susceptibility of devices (see Figure 2-2). However, other models are available which might more accurately represent what happens in specific cases.

3. THE EFFECTS OF ESD

STATIC DISCHARGING

Electrostatic discharge occurs either when a charged conductor contacts another conductor at a different potential as an electrical ground or a resistive path to ground, or when an insulator or conductor is so highly charged that its electric field breaks down the medium separating the charged object from another object at a different voltage. The rate of discharge is determined principally by the path resistance. When an electronic device is in the discharge path, it can be damaged either by excessive internal heating or by dielectric breakdown. Breakdown can also be caused by potentials induced across dielectrics in strong electric fields.

FAILURES CAUSED BY ELECTROSTATIC DISCHARGE

There are primarily three ESD events that can cause device degradation or failure:

1. Should an electrostatically charged person touch a device, a transfer or discharge of some of the energy stored in the person's body occurs either to the device or through the device to ground. (See Figure 3-1.)
2. Should a device or package itself acquire a charge on its lead frame or other conductive parts, it may discharge to ground through one of its leads. (See Figure 3-1.)
3. Should the gate structure of a device be inserted into a field, it can develop a nonzero potential. If a device is grounded while in the influence of an electrostatic field, it will experience a surge of current through the grounded lead. (See Figure 3-1.)

Static electricity is charge at rest on an object, caused by a transfer of electrons. This charge can be either negative, if there is an excess of electrons on a body, or positive, if there is a deficiency of electrons on that body. One method of electron transfer occurs when two contacting bodies are rubbed together and then separated—a phenomenon known as the triboelectric effect.

TRIBOELECTRIC CHARGING

Anytime two materials come in contact with some movement between the two and are then separated quickly, an equal and opposite charge is developed on the surfaces. This also occurs with gases flowing across a surface. When one material is a conductor, it retains the charge until grounded; this is considered a mobile charge. If the material is an insulator, the charge is immobile, and it remains until neutralized. This frictional charging is called triboelectric charging.

For example, as a warm, moist air mass collides with a cold, dry air mass, the warm air rises and the cold air sinks. The movement of one air mass against the other causes a positive charge to build up on clouds that are formed. When the charge on the clouds get big enough, a discharge takes place that we see as a bolt of lightning.

This same process causes a static charge on our body anytime we move. Our skin and clothing rub the air and, as we walk across a carpet, our feet separate from the carpet. The rubbing and separating actions cause a static charge to build up on our body and clothing. As long as we do not touch a conductive surface, the charge will remain static. As we approach a conductive surface such as a doorknob, however, at some point—depending on the amount of charge built up—the charge exceeds the insulating ability of the air, and a spark will jump from our body to the doorknob.

Triboelectric charging is the process that generates the charge in both cases cited. So long as the charge does not move to ground, the charge is called a static electrical charge. The contact and separation of two surfaces causes an interchange of electrons to take place. Some materials give up electrons very easily and become positively charged, while others accept extra electrons easily and become negatively charged. This property is inherent in each material.

A material near the top of Table 3-1 is charged positively when contacted by a material below it. For example, if you comb your hair with a hard rubber comb, your hair will become positively charged and the comb negatively charged. When a glass rod is rubbed with a piece of rabbit fur, the fur will be positively charged and the glass rod nega-

Table 3-1. The Triboelectric Series.

TRIBOELECTRIC	CHARGE
Air	+
Human hands	↑
Asbestos	
Rabbit fur	
Glass	
Mica	
Human hair	
Nylon	
Wood	
Fur	
Lead	
Silk	
Aluminum	
Paper	
Cotton (reference point)	
Steel	
Wood	
Amber	
Sealing wax	
Hard rubber	
Nickel copper	
Brass silver	
Gold platinum	
Sulfur	
Acetate rayon	
Polyester	
Celluloid	
Orlon	
Saran	
Polyurethane	
Polyethylene	
Polypropylene	
PVC (vinyl)	
KEL-F (CTFE)	
Silicon	↓
Teflon	−

tively charged. The further apart two materials are in the series, the greater the charge.

The triboelectric series has little to do with the magnitude of charge that is generated. The magnitude of charge generation is influenced by such factors as contact pressure, the speeds of separation or rubbing, contact area, lubricity, and ambient conditions.

ELECTROMAGNETIC INDUCTION

Direct contact between objects is not the only way a static charge can be generated. A statically charged object, when brought in close proximity with an uncharged object, can induce a charge on the formerly neutral object. This is known as an induced charge.

If a conductor is placed close to a negative charged plastic box, electrons in the conductor will migrate away from the surface closest to the box, leaving this surface with a net positive charge. If an uncharged nonconductor is brought near the same box, electrons will not flow, but the closest surface will become charged as the molecules form a dipole.

Once a material is charged by the triboelectric effect or some other means, the charged material has an electromagnetic field surrounding it. This field is similar to the field around a common magnet. It surrounds the statically charged material. It is just as real as the magnetic field around a magnet and has identical properties.

When a conductor is moved through a field, or the field is moved near a conductor, an electrical current is induced in the conductor. This process is called electromagnetic induction. Static-sensitive devices are a combination of conductors and insulators. Simply walking across a carpet can build up a charge of 20,000 volts, and as a worker approaches a static-sensitive device without touching it, the static field surrounding his body would pass through the conductors of the device. This induced current could lead to different voltages across the circuits in the device, causing both voltage-induced or current-induced failures, or degradation of the device.

The important things to remember about static fields and induced charges are:

1. Static fields are invisible, but they exist around all charged materials.
2. No physical contact between the charged material and the conductor is necessary to induce a charge.

3. So long as the field moves near a conductor, or the conductor moves near the field, an induced current will flow in the conductor.
4. The amount of current that will flow depends primarily on the strength of the field and the speed of the movement.

ELECTROMAGNETIC PULSE

Whenever a charged object discharges and a spark occurs, the spark generates a separate traveling electromagnetic field. This separate field is identical to the static field, except that it has its own motion, magnitude, and duration. This traveling field can induce currents in both stationary and moving conductors, thus causing the same kinds of damage as induced charges. (See Figurc 3-1.)

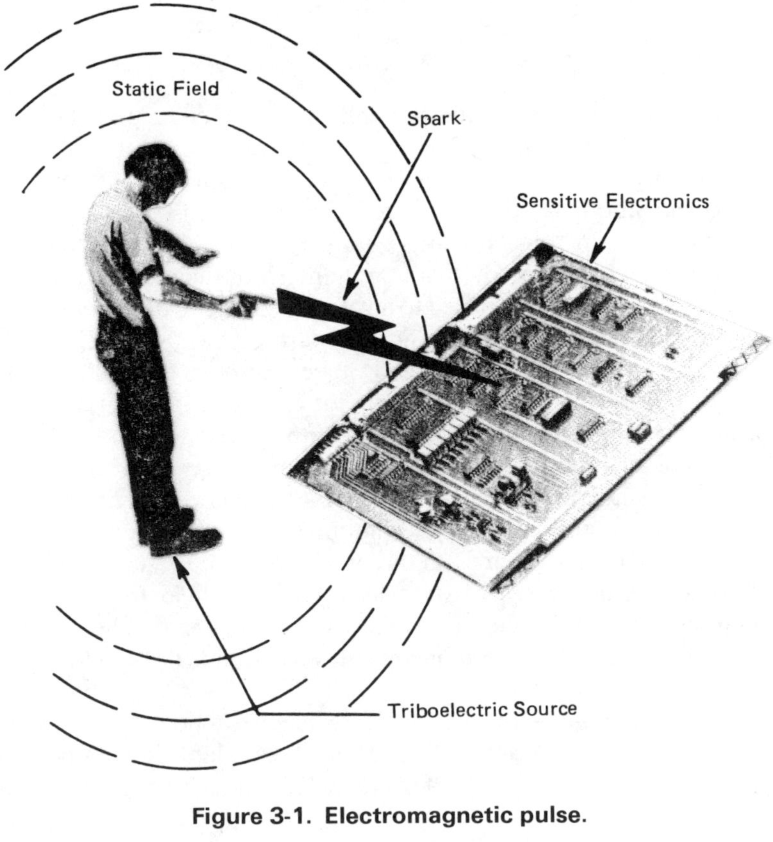

Figure 3-1. Electromagnetic pulse.

Every time a spark occurs, a static field is generated. When lightning strikes, a tremendous field is generated and though you may have a good AM radio with a network of filters to filter out "static noise," your AM radio will pick up this signal, amplify it, and drown out the signal of the radio station you are listening to.

At the instant the charge begins to move, voltage rises very rapidly over a very short period of time until it reaches a peak, at which point the movement of the charge to ground limits a further rise in voltage, so that the voltage begins to drop back down to zero. During the rise time (from zero to peak), electronic devices which are voltage-sensitive can be damaged, if the peak voltage is more than the device can stand. During the decay time (from peak back down to zero), electronic devices which are power-sensitive can be destroyed, if the time it takes to return to zero is too long.

POLARIZATION

If a conductor is exposed to an electromagnetic field, and the field and the conductor are both stationary, a phenomenon known as polarization takes place. When any material is charged, it emits an electromagnetic field. If the charged object is an insulator such as a styrene coffee cup, it will hold the charge for a long period of time. As long as it remains charged, it will emit an electromagnetic field.

When a conductor is exposed to an electromagnetic field, some of its electrons become mobile and move toward or away from the field, depending on whether the charged object emitting the field is positively or negatively charged. Assume, for example, that you have just finished drinking coffee from a styrene cup, and you set it down on a workbench near a sensitive electronic device. As the cup moves through the air prior to being placed on the bench, the triboelectric effect of moving through the air, combined with your handling of the cup, creates a positive charge on the surface of the cup. As the charged cup passes near the sensitive device, the static field from the cup sweeps across that device, inducing a current flow in it. The induced current in the device can damage or destroy it, nor does the process stop there.

When the charged cup with its electromagnetic field comes to rest near the device, polarization of the device takes place. Remember, the cup need not touch the device for this to happen. The cup and device are now close together, and both are stationary. Once the device gets into the electromagnetic field of the cup, mobile electrons in the conductors move toward the charged cup, leaving the opposite end of the device deficient in electrons (positively charged). The electrons in the device behave according to the law of physics that says, "Like charges repel; unlike charges attract." However, at this point the device is not charged; it is polarized. (See Figure 3-2.)

Should the device be picked up, moved, and placed onto a grounded surface, it would become charged the instant it was touched and moved out of the field. Then the instant it touched ground, it would be discharged.

This instantaneous flow to ground causes such rapid movement of electrons in the device that heat is generated; if the charge created is large enough, the discharge will generate enough heat to actually melt parts of the device. The current induced in the device when the cup was originally moved to its resting place may have only partially damaged the device. However, when the charge created by polarization is discharged, the weakened device can be seriously damaged.

This is not at all an unrealistic situation. Before static electrical discharges and associated field effects were identified as the cause of

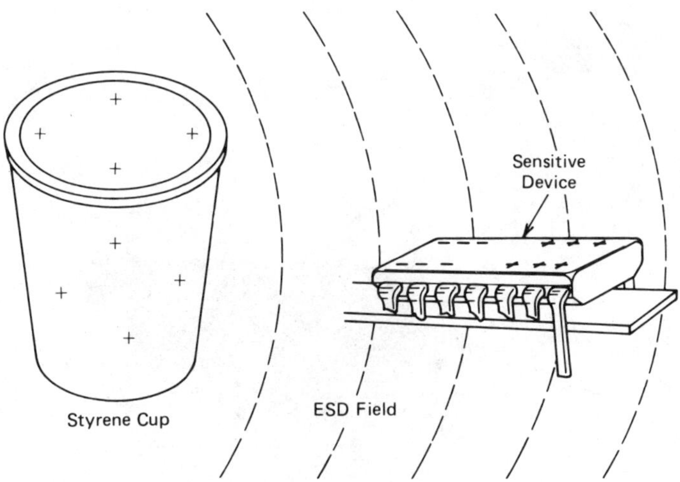

Figure 3-2. Polarization.

damage to sensitive devices, it happened every day. It may not have been a coffee cup that was involved, but perhaps the charge from a chair or a nonconductive paperwork holder, or even from a roll of tape in a dispenser on the workbench.

FREQUENCY COUPLING

Certain radio frequencies, such as those related to radar and microwave transmitters, can be picked up by electronic components and be destroyed or damaged through a process called frequency coupling. You have probably been in an auditorium or a large meeting room where a microphone hooked up to an amplifier with speakers began producing a sharp, ear-piercing screech when a person started to talk, or when some other sound was picked up by the microphone. This is frequency coupling.

Sound waves picked up by the microphone are converted to electrical pulses. The magnitude of the pulses is boosted by the amplifier and passed on to the speaker assembly, where they are converted back to sound waves. If the speaker assembly is placed in a position where these sound waves are picked up by the microphone, and the distance

Figure 3-3. An isolated device exposed to ESD; it retains the charge.

from speaker to microphone and the frequency of the sound waves are just right, another set of electrical pulses is generated by the microphone and the entire cycle keeps repeating. Each time the cycle is repeated, the frequency raises the pitch of the sound to an ear-splitting screech.

During the transport of sensitive devices by rail or truck, the vehicle may pass through areas where damaging frequencies exist. Radar units at commercial airports or local TV stations for monitoring the weather, as well as military radar used on land and sea and in the air, emit frequencies that are potentially damaging. In addition, microwave transmitters used by the telephone company blanket the country with microwave transmissions.

If the distance and the frequency are right, sensitive devices in a moving truck can be damaged as they pass through these invisible wave forms. Actual cases of computer fire-control and guidance systems failures on naval vessels and in military aircraft have been traced to

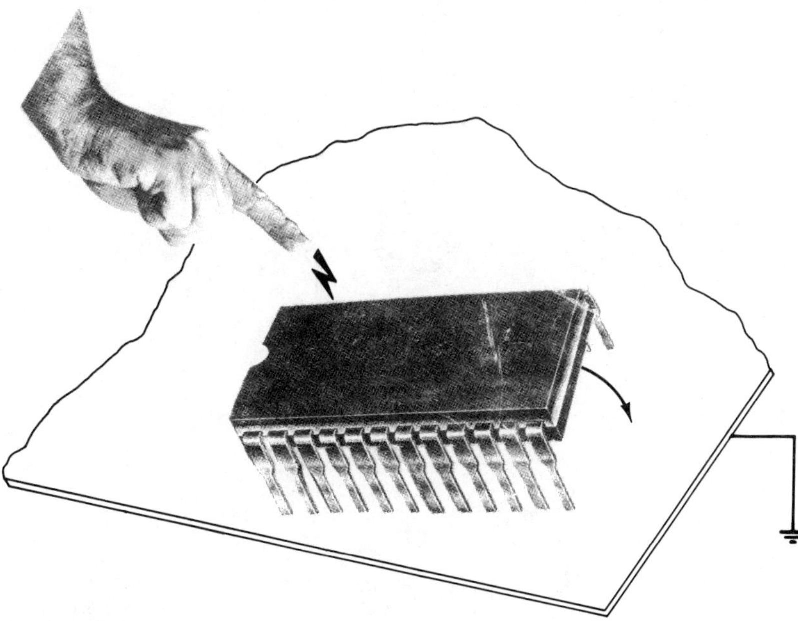

Figure 3-4. A device placed on an ESD mat while exposed to ESD drains to ground.

frequency coupling damage caused by the search radar unit of an adjacent ship or aircraft.

Operational amplifiers (OP-AMPS) with unprotected MOS capacitors, microwave and ultra-high frequency semiconductors, and microcircuits are particularly sensitive to damage from frequency coupling caused by radar and microwave transmissions. These wave forms will pass right through insulators such as plain polyethylene or glass. However, they can be stopped by a barrier material such as aluminum foil.

4. CONTROLLED ENVIRONMENTS

Charges residing on nonconductors do not, by definition, allow electron flow and cannot be drained off by grounding, so other means must be used to eliminate static charges which build up on them. Elimination of nonconductive materials from the work area is one way of controlling static electricity. While removing all nonconductive materials from the work area is commonly attempted, the total elimination of these materials is difficult and requires constant attention. For all practical purposes, if potentials in the work environment can be maintained at less than 100 volts, static problems should not exist.

Component destruction is costly and time-consuming, but is usually caught before the circuit board or device leaves the factory. Expensive and elusive problems result from devices that are only slightly altered or degraded after exposure to static electricity. They might function within specifications during the final quality-assurance check, but their service life can be severely shortened.

Fortunately, the problems associated with static charges in the electronics environment can be controlled. Programs can be implemented to increase the level of static awareness in all personnel who handle static-sensitive devices. Discussion of static safe operational and handling techniques are important since, as shown in Table 1-1, sensitive devices are damaged by potentials that the discharging person cannot even feel (less than 3,000 volts). There is no indication of the problem until a malfunctioning part is located at the next test point or—even worse—in the field.

Educated personnel must be provided with the proper equipment to implement these techniques of static control. At the workbench this includes conductive table mats, conductive wrist straps, conductive floor mats, and ionized air.

Static electricity is an excess or deficiency of electrons on a surface. The total number of deficient or excess electrons determines the charge on that surface. A surface exhibiting an excess of electrons is charged negatively, while one with an electron deficiency is charged positively. The basic concept of static protection for electronic components is the prevention of static buildup where possible, and the quick, reliable removal of existing charges. The means by which these charges are removed depends upon whether the charged object is a conductor or a nonconductor (insulator). These two types of materials coexist throughout the electronics manufacturing environment. The selection of proper static control equipment is critical; wrong decisions can generate more problems than solutions.

BASIC GROUNDING EQUIPMENT

Grounding equipment comes in many forms and types. Essentially, it includes grounded conductive floor mats for operators to stand on; grounded conductive work surfaces which can be an integral part of the work area or be placed on top of existing nonconductive surfaces; and personal wrist straps, which are grounding devices to ground the personnel. These types of devices should have a current-limiting resistor to prevent hazardous shocks to personnel, should they come in contact with a live circuit. In all grounding devices, surfaces and mats should be made of conductive material. Likewise, they must have an effective grounding wire connected to a good earth ground. (See Figure 4-1.)

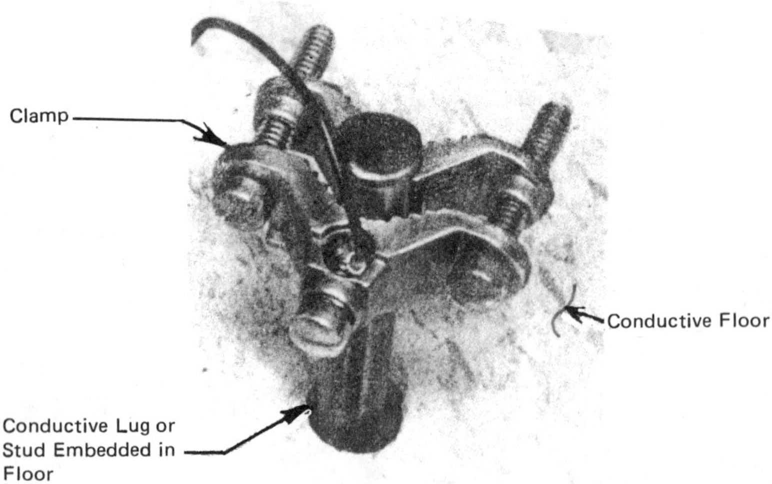

Figure 4-1. Typical static control ground connection.

IONIZERS

Ionizers are manufactured in a wide variety of configurations and systems. Functionally, their purpose is to neutralize static electricity already in existence or to prevent the development of static charges. Operationally, ionizers operate by one of two methods: radioactive ion generation or high-voltage ion generation. The equipment for the former is leased rather than sold, the radioactive element must be returned to the source annually for inspection and recharging. High-voltage units, on the other hand, are usually sold outright.

In selecting ionizing equipment, you should be specifically concerned about the following:

- RF signal generation shielding.
- Personnel protection in the form of current-limiting and mechanical protection on the ionizing points.
- Magnetically controlled power sources.
- Low maintenance and ease of maintenance when maintenance is, in fact, required.
- Factory service.
- Low ozone generation.

THE USE OF IONIZED AIR TO NEUTRALIZE IMMOBILE STATIC CHARGES

Because of the difficulty in controlling and eliminating charged insulators from the electronic manufacturing work environment, air ionizers are often used to neutralize these charges. Since positively or negatively charged surfaces will attract ions of the opposite polarity, an air stream containing both positive and negative ions will neutralize the charged surface. Once the surface is neutralized, it remains so as long as the ion stream is reasonably balanced.

Ionized air used in electronic manufacturing is generated by either alpha-particle bombardment of the air or by corona discharge. Radioactive neutralizers ionize the air by bombarding it with alpha particles emitted from Polonium 210. Corona discharge ionization places high voltage on a sharp pointed pin located in close proximity to an electrical ground. The air is ionized as the electrical energy attempts to arc across the gap. When a single high-voltage polarity is applied to the pointed pin, the system will generate ions of a single polarity. However, voltages of different polarities must be applied in order to generate both positive and negative ions. Since both negatively and positively charged surfaces are encountered, DC ionizers usually have two pins. One pin is at a negative potential and the other at a positive.

AC ionizers, on the other hand, apply a high-voltage signal of alternating polarities, so that part of the time the unit is generating positive ions and part of the time, negative ions.

Several criticisms of corona discharge and radioactive ionizers have been voiced. Radioactive ionizers require rental of the radioactive element. Some point out that this type of unit, contains radioactive materials, but the alpha particles are considered safe. The large mass of the emitted alpha particles can be stopped by shields as thin as a sheet of paper. On the other hand, corona discharge ionizers will generate some ozone, but a properly designed ionizer will be within limits that are considered safe.

Another problem is that equivalent voltages of opposite polarity generate slightly different amounts of positive and negative ions. This imbalance is corrected with an altered wave form, as in the case of AC ionizers, or by utilizing slightly different voltages for dual-pin DC units.

Both radiation and corona discharge ionizers will lose efficiency with time. Radioactive materials have a half-life, which means that in time they emit fewer alpha particles. Corona discharge units can lose some efficiency through erosion of the ionizing tip.

Air ionizers come in a variety of configurations: bars, air guns, portable bench-top units with blowers, laminar-flow hoods, or entire room systems.

Substantial savings are realized by eliminating the need to keep a facility at 40% relative humidity. An additional benefit of using ionized air is that contamination control is improved.

Some words of caution: the coverage provided by ionization systems varies. Some of the bench-top units use fans and blowers to increase the coverage area. One will find various claims about which fan-system ionization techniques give the best coverage. But is a slight difference in ion balance really significant?

Judge selections on a strong technical base. If you choose to use ionized air, remember that it must be monitored to assure that it is operational. Do not assume that, just because a blower is working, the system is generating ions. Power might have been lost and the system may be generating no ions at all, or in the case of a DC electrical unit, it may be generating ions of a single polarity.

Ionization can be achieved by three methods: through induction, by the use of radioactive alpha particles, and with high voltage in close proximity to ground. Not all three are practical for the electronics industry. For example, ionization by induction is impractical for controlling static electricity of low potential, since a threshold (or starting) voltage is required to start the process, so that the static electricity will never be reduced to a zero. Since most static electricity levels are below the threshold voltage in the electronics industry, this is not very practical.

Basically, there are two types of ionizers available for use in the electronics manufacturing environment: radioactive and electrical. Both types are manufactured in the form of blowers, guns, nozzles, and a variety of special shapes. Each form can be supplied with or without an attachment for air assistance.

RADIOACTIVE ALPHA PARTICLES

Ionization by utilizing radioactive material is another method. Polonium 210 is a common material used for this purpose. The Polonium 210 emits alpha particles that bombard the air, creating both negative and positive ions. It is an excellent source of ionization because, for every positive ion it generates, it generates a negative ion simultaneously. (See Figure 4-2.) The ability of Polonium 210 to emit alpha particles diminishes rapidly within the first 200 days or less. From then on it

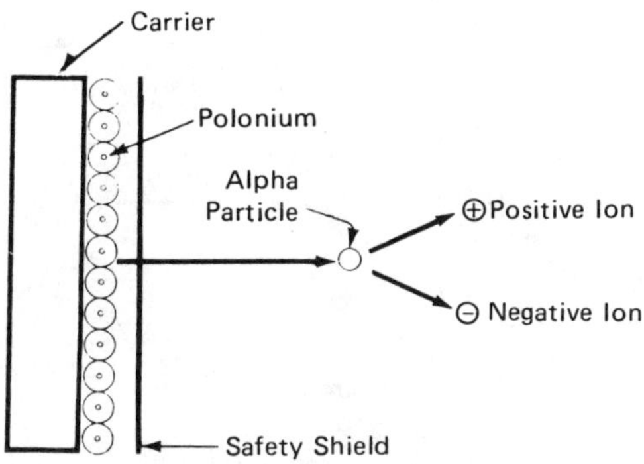

Figure 4-2. Alpha particle bombarding air molecules to create positive and negative ions.

diminishes very slowly and will last beyond the year required to return the unit for leak testing.

Radioactive ionizers are very effective. Moreover, they have the capability of being used under explosive atmospheric conditions without fear of causing an explosion. This is due to the fact that no high energy, such as electricity, is required to produce the ionization.

These ionizing units can only be leased from manufacturers and are licensed by the Atomic Energy Commission. By law, they must be returned to the manufacturer yearly to be leak-tested. To prevent leaking, the protective beads should not be disturbed. If you are using these ionizers and they stop functioning, like any other piece of equipment they should be returned to the manufacturer.

Radioactive types are efficient ionizers and have the advantage of requiring no wiring or related power equipment. However, since they must be leased and replaced each year, they require extensive paperwork and shipping expenses. On the other hand, since these units are replaced yearly, the cost of equipment maintenance is eliminated.

The simplest and most cost-effective method of generating ionized air is by use of high voltage on a sharp point which is placed in close proximity to ground. (See Figure 4-3.) High-voltage ionizers function on the principle that high electrical energy concentrated on a sharp point, in close proximity to an electrical ground, will break the air down by ionization as the energy attempts to arc the gap between the point and ground. When the ionizer is designed properly, the energy will never arc because the gap is too wide. But if this gap is too great, very poor ionization will occur and the ionizer may be unable to control your static problems.

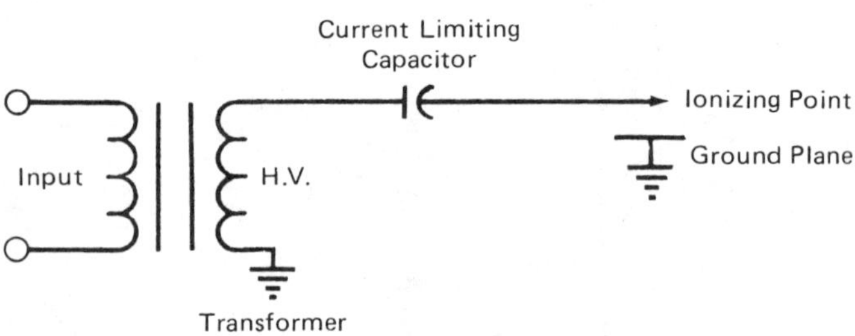

Figure 4-3. Circuit diagram for an electrical ionizer.

To control static electricity, both positive and negative ions must be available, because both polarities of static electricity are being generated in your plant. Therefore a positive charge will require negative ions to be neutralized, and a negative charge will require positive ions to be neutralized. This is because like charges repel each other, while unlike charges attract each other.

When the alternating high voltage passes through its cycle, it will generate positive ions during one alternation and negative ions during the other. Starting from ground level, the voltage on the point will reach point A as shown in Figure 4-4, and positive ions will start to be generated. As the voltage continues through its signal, it will reach point B and ionization will cease. Continuing, it will reach point C and ionization will again begin. However, the voltage will now cause negative ions to be generated, because the voltage gradient is built up in the opposite direction from ground.

As the electronics industry becomes more sophisticated, greater requirements are placed on manufacturers of ionizing equipment. Sophisticated test equipment is now being utilized that cannot tolerate RF signals. Make sure this is not a problem, if ionizers are being used in the vicinity of test equipment or special control circuits. Complete shielding against RF signals may be necessary.

All electrical ionizers will produce ozone. The maximum allowable concentration, by law, is 0.02 part per million. Most electrical ionizers operate well below this limit. However, there are ionizers on the market with questionable ozone output. Care should be exercised when purchasing this equipment.

Since OSHA must also be considered, all ionizing points should be

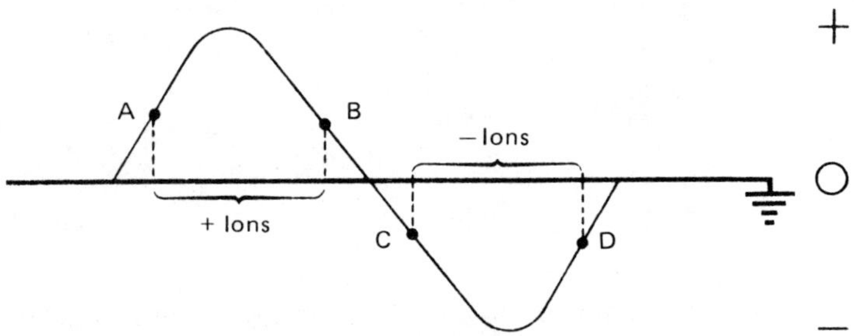

Figure 4-4. The ionization cycle.

mechanically protected to prevent operators from hurting themselves, should they grab the ionizer or fall against it.

Keep in mind that ionizing equipment must be tested periodically to assure that it is functioning properly and providing you with the protection your static-sensitive devices require.

Electrically powered ionizers can be purchased outright and are a one-time investment. If you purchase the right electrical ionizers, you can get many years of trouble-free use from this type of equipment. Only two parts of an electrical ionizer can wear out—the fans and the ionizing points. If the device you purchase utilizes an inexpensive fan, it will require maintenance or replacement in a short time. So check the quality of the fan in the particular unit you are considering.

If the ionizing points are manufactured from a mild steel, they will lose their sharpness through erosion, and the efficiency of the ionizer will begin to decline. Sharpness is required to produce the proper amount of ionization.

Air-assist static neutralizers are manufactured in two basic forms. One type utilizes high-pressure, low-volume ionizing air. These take the form of air guns, nozzles, and special forms of air jets. They are designed primarily for cleaning. Very small dust particles can be attached to an electronic assembly by electrostatic forces that are greater than the force created by the blast from a conventional air gun. If one uses an ionizing air gun, the electrostatic forces will be neutralized,

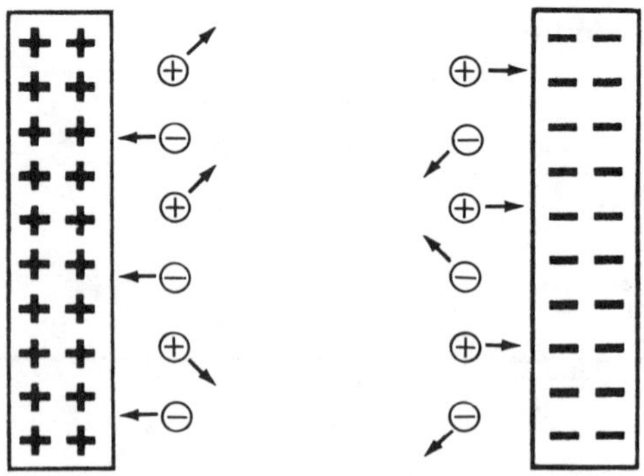

Figure 4-5. Neutralization by ionization.

while the part will be cleaned and neutralized. Nitrogen also can be ionized with these ionizing air guns.

A second type of air-assist ionizer is the low-pressure, high-volume type, designed for neutralizing static electricity at great distances. This type of unit is utilized in assembly areas. These blowers will control static electricity at a range of six feet. They control static electricity not only on nonconductors but on an operator's clothing—an item which is often overlooked.

Ionizers that are furnished without an air-assistance attachment can be used to control static electricity on wave solder machines. An example is a static-neutralizing bar that has a maximum range of one inch from the charge surface. This means that the material or device to be neutralized must pass within one inch of the working surface of the static neutralizer.

When choosing ionizing equipment, several factors must be considered. Safety standards must be considered, including ozone output. Size of a unit is a factor, because production space is expensive. Ionizing air blowers above 100 cfm can create discomfort to the operators. When this happens, there is a tendency to turn the units off, or direct the ionized air stream away from the working area. Ionizing air blowers with adjustable fans can resolve this problem.

When evaluating ionizing equipment, keep the electrostatic field in mind. If this field is disturbed or collapsed, no ionizer will neutralize static electricity effectively. If a tote tray laden with static electricity is placed on a table in front of an ionizer and the ionizer is then turned on, the static electricity on the sides will be neutralized but the bottom surface, contacting the table, will not, because it was placed against the tabletop before the ionizer was turned on and the electrostatic field collapsed. The ionizer must be operating before the tote tray is taken into the area. Teach your operators to turn on the ionizing equipment first, before beginning any other work.

Adding humidity is another method of providing a moisture layer to the nonconductive plastic. However, to be effective, the humidity must be raised to the point where the perspiration produced by the human body cannot evaporate, so that operators will be very uncomfortable. Adding moisture by raising humidity is a temporary solution to the problem: since there are no antistatic agents involved to retain or store the moisture layer, it will evaporate upon leaving the high-humidity area. Furthermore, adding humidity to a room creates other problems, since the moisture cannot distinguish between conductors and insulators. All your equipment will be laden with moisture and problems can develop, ranging from the rusting of expensive equipment to changing the parameters of test devices. Extreme care should be used when adding humidity to production areas.

During the summer, humidity is high and nature's moisture layer makes items such as a nylon rug antistatic.

TESTING AND MONITORING IONIZERS

Figure 4-6 shows a typical test circuit for electrical ionizers. Note the loop in the circuit that is opened by the air gap between the ionizer and the metal plate. If this gap is closed, current will flow in the loop. The current flow can be measured with the microammeter in the circuit. Since ionized air is conductive, the gap can be closed with ionized air; the current flow will actually be ionization current. The higher the ionization current, the more efficient the ionizing device, because it is generating more ions. It is a relatively simple test, and can be repeated to provide reliable information.

However, the size of the metal plate must be kept constant (usually 12″ × 12″). The air gap must be kept constant (one half inch). The high voltage A.C. source to the ionizers must be kept constant (depends on

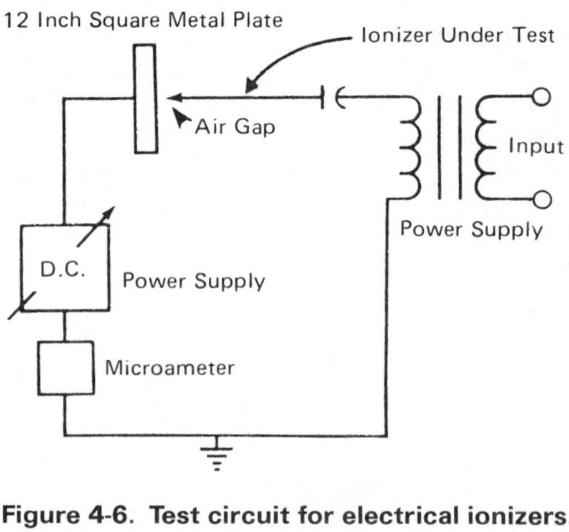

Figure 4-6. Test circuit for electrical ionizers.

the design of the ionizer), as must the high voltage D.C. source also. Microampere readings should be recorded at 500 V.D.C., 1,000 V.D.C., 2,500 V.D.C., 5,000 V.D.C., and 10,000 V.D.C. The test should be repeated with both positive and negative voltage applied to the metal plate, in order to reproduce the static electricity the ionizer is trying to neutralize.

Figure 4-7 shows the typical curve generated by electrical ionizers. If you would like to simplify the test, just pick any two values of high voltage D.C. and maintain the rest of the constants; the test will be reliable. If the ionization current starts to drop off, the ionizer needs cleaning or repair.

FLOOR MANAGEMENT ALTERNATIVES

There are a number of alternatives for effective static control for floors: floor mats, topical treatments, or floor coverings.

Floor Mats

Probably the most familiar is the floor mat, available in four major types. First there is the black, carbon-filled, conductive mat. Either rubber, vinyl, or polyolefin, it has an electrical resistance of from 10^3 to 10^5 ohms per square. Second is the colored, cushioned laminate with an electrical resistance of from 10^6 to 10^9 ohms per square. Third is the

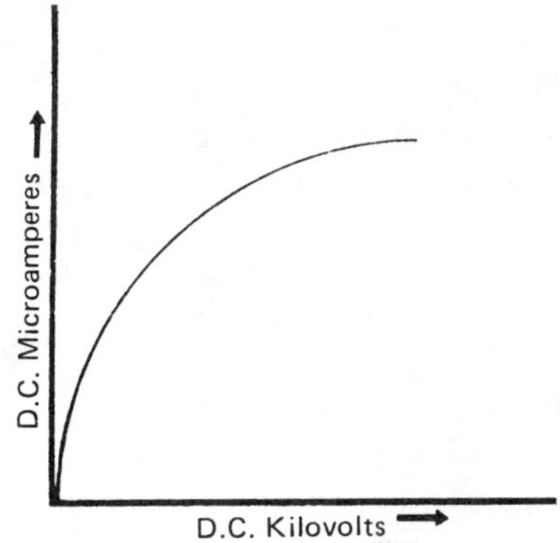

Figure 4-7. Typical curve of test results.

antistatic mat, typically a rigid polyolefin or ABS, with an electrical resistance of from 10^{10} to 10^{13} ohms per square. The fourth type is the rigid laminate with an electrical resistance of from 10^3 to 10^8 ohms per square. All mats are loose-lay materials usually grounded through a one-megohm current-limiting resistor. Static decay rates per Federal Test Method 4046 range from 0.01 second to more than 10.0 seconds.

Mats in general—and particularly the carbon-filled type—have proven their effectiveness in controlling static at localized work areas. In addition, mats are relatively easy to install and can provide immediate static protection. But floor mats have some limitations. They provide only localized protection, expanding static control slightly beyond the boundaries of the wrist strap. Supervisors or other employees who approach work areas must make direct contact with the mat to effectively control static problems. Also, some mats are humidity-dependent, limiting their effectiveness. Many tend to curl and must be taped down to hold them in place and to reduce tripping hazards. In addition, maintenance personnel must either move and replace the mats or attempt to clean around them. Some mats, particularly the carbon-filled ones, are not applicable for clean room situations, because of a potential for particulate contamination.

Topical Treatments

Topical treatments are those applied to existing floors intended to render them static-safe. In general, these treatments take three forms: conductive paints or coatings, topical antistats, and conductive or antistatic floor finishes.

Conductive paints or coatings. Typically, these are latex paints or acrylic coatings applied to the floor with a brush or roller. They range in resistance from 5×10^3 to more than 5×10^5 ohms per square. In general, their static performance is not well documented, although static decay times are less than 0.1 second per Federal Test Method 4046.

The major advantage of these materials is that they can provide environmental rather than localized static protection and can be readily applied in existing facilities. On the other hand, conductive paints and coatings have several limitations. They are usually carbon loaded and messy to work with. They are subject to wear, flaking, and chipping and are not applicable for clean rooms. They require frequent reapplication and, over a period of time, can become expensive compared to some other alternatives.

Topical Antistats. A more familiar product in the electronics industry is the topical antistat, a liquid consisting of two components: an antistatic mechanism and a carrier to transport it.

Topical antistats function in three ways. First, they reduce the coefficient of friction to help minimize frictional static charge generation. Second, they increase surface conductivity to help dissipate a charge. Third, they interact with environmental factors to neutralize generated charges. Some absorb moisture from the environment to help create a conductive layer on the substrate. Others rely on an ionic interaction between the environment and the antistat to help neutralize the charge. Because most antistats are diluted for use with water, they are easily applied to floors in the normal cleaning process. In addition, they provide environmental static protection. They are generally less expensive per application than conductive paints or coatings.

Topical antistats also have their limitations. First, because they are not principally intended for floors, scuff resistance and glossy appearance are sacrificed. Second, because most are soluble in both water and solvents, it's easy to accidentally remove the antistat from the floor. Third, continual reapplication is necessary, usually more frequently than the manufacturer suggests. Fourth, there seems to be variation in the features, quality, and performance among the various brands. For example, some are hygroscopic and their performance is humidity-dependent. Some antistats may contaminate clean room environments, while others may be acceptable for these applications. Some may cause skin or eye irritation when applied.

Proper management of floor surfaces helps provide environmental static protection, deals with causes as well as symptoms of static, and complements other elements of a static control program.

Conductive or Antistatic Floor Finishes. These finishes have recently started to find their way into the electronics industry. In contrast to topical antistats, which have many different end uses, floor finishes have been designed specifically for resilient floors, so that they have advantages in scuff resistance and appearance.

Like topical antistats, conductive or antistatic floor finishes offer environmental static protection, ease of application, and a relatively modest cost per application. But they also have their limitations. The first is their static performance: one recent study has indicated that the products tested perform no better than untreated floors. Second, some products also appear to be humidity-dependent, limiting their effectiveness. Third, one doesn't know how effective the static control is at any point in time. Regular monitoring is required and continual reapplication is necessary. Because of this, the apparent low initial costs can be misleading.

Floor Coverings

Static control floor coverings are not a new concept, although they are fairly new to the electronics industry. Conductive vinyl floor coverings have been used for more than 35 years in hospital surgeries. Antistatic carpeting, however, is a more recent product, developed specifically to control static in computer rooms.

Vinyl Floor Coverings. Conductive vinyl flooring is usually available as 12″ × 12″ tile, 36″ × 36″ sections, or sheet goods. The sheet goods are generally carbon-filled, tend to be black, and are physically similar to conductive mats. The tile is installed with a conductive adhesive that provides lateral conductivity. Together, the tile and adhesive create a pathway of moderate electrical conductivity to ground.

The biggest advantage of conductive vinyl flooring is that it provides complete environmental protection without the need for additional floor mats or topical treatments. It requires no special maintenance to retain its conductivity, and creates an attractive and pleasant work environment. Conductive tile also may be used in clean rooms and on raised flooring. A disadvantage of conductive flooring is its high initial cost. Life-cycle costing, however, shows conductive flooring to be cost-effective.

Antistatic Carpeting. In the past few years, antistatic carpet has been developed for use in computer rooms and similar less sensitive environments. Although you might want to consider carpet because

Table 4-1. Floor Surface Alternatives: Benefits and Limitations.

PRODUCT	BENEFITS	LIMITATIONS
Floor mats	Static performance, particularly the conductive mats Ease of installation/use Moderate useful life	Localized protection Curling Maintenance Periodic replacement Some are humidity-dependent Carbon-filled type not clean-room-applicable
Paints/ coatings	Static performance Application to existing floors Environmental protection	Usually carbon-loaded Frequent reapplication owing to wear, flaking, chipping Not clean-room-applicable
Topical antistats	Environmental protection Application to existing floors Cost per application	Limited static performance Continued frequent reapplication required Variation from brand to brand Some may cause skin/eye irritation Some are not clean-room-applicable Some are humidity-dependent Easy to accidentally remove treatment Lack of gloss and scuff resistance Dependence upon maintenance department for static control
Conductive/ antistatic floor finishes	Application to existing floors Environmental protection Scuff resistance and gloss	Limited static performance Some are humidity-affected Continued frequent reapplication required Hard to know when it's ineffective Dependence upon maintenance department for static control
Conductive vinyl flooring	Static performance Environmental protection Durability/permanence Attractiveness Clean-room-applicable (tile only) Can be used on raised floors (tile only)	Initial cost Installation inconvenience in existing areas Sheet products not applicable for clean rooms or raised floors

Table 4-1. Floor Surface Alternatives: Benefits and Limitations. ***(continued)***

PRODUCT	BENEFITS	LIMITATIONS
Antistatic vinyl flooring	Environmental protection Durability/permanence Clean-room-applicable	Static performance Initial cost Installation inconvenience in existing areas
Antistatic carpeting	Environmental protection Acoustics Aesthetics	Static performance Maintenance Wear/durability Chemical/solder resistance Installation inconvenience in existing areas

of its acoustical or aesthetic properties, it has poorer static control than many alternatives. In addition, most carpeting is not designed to withstand the chemicals, solder, and general wear and tear of many electronics manufacturing environments.

Conductive Floors. In response to the rapidly growing industry awareness of the need for methods and materials to control ESD, conductive flooring is receiving increasing attention and is being incorporated into the ESD control programs. To much of the industry, conductive flooring still remains relatively unknown.

Historically, conductive floors have been used to reduce the risk of explosions or fires caused by an electrostatic spark discharge in hospital surgeries or other flammable and hazardous environments. Today, conductive floors are finding applications in data processing and other computer facilities, in electronics repair and maintenance shops, in installations of highly sensitive and sophisticated electronic equipment, and in electronics manufacturing, assembly, and test facilities.

Vinyl Conductive Floors. This flooring is the most commonly used material. It has a proven record of performance, attractive appearance, ease of maintenance and installation, chemical resistance, underfoot comfort, and cost-effectiveness.

These specialty floors are carefully engineered systems comprising conductive vinyl tile and epoxy adhesive. The tile contains conductive elements of the carbon family that are heat- and pressure-fused and permanently sealed into a homogenous floor tile. The conductive elements are randomly distributed throughout the tile's surface and thickness. The conductive elements provide through-tile conductivity, while the conductive adhesive provides lateral tile-to-tile conductive continuity creating a pathway of moderate electrical conductivity to ground.

Vinyl conductive floors perform two major functions. They control static charges on personnel and equipment by minimizing the generation and accumulation of charges caused by contact separation, and they dissipate static charges from personnel or objects that come in contact with the floor.

Benefits and Limitations of Floor Coverings

Aside from the natural function of static control, there are other important benefits in using conductive flooring in the electronics environment. First, because conductive floors can be installed throughout an entire manufacturing area, static protection is not limited to individual work stations. Personnel moving throughout the area can approach work stations in relative safety, since only minimal static charges are generated as they walk. Conductive flooring also provides protection in instances when an individual forgets to attach a wrist strap or to step on a conductive mat.

Because permanently installed vinyl conductive floors provide a continuous smooth surface, they are easier to maintain than conductive mats that must be moved every time the floor is cleaned. They also eliminate tripping hazards. Natural resilience also combats leg and foot fatigue in employees who work on their feet much of the day.

Conductive floors have limitations as well. One of the principal limitations is that conductive flooring by itself is not the whole answer to ESD control. But considered as part of the overall ESD program and used in conjunction with other conventional products, it is a first step in achieving the level of required protection.

Thick rubber- or crepe-soled shoes are excellent insulators, completely defeating the purpose of conductive floors, and should be restricted from manufacturing areas. Leather- or composition-soled shoes may be suitable in some situations, depending upon the sensitivity of the parts and assemblies being handled. In the most highly critical areas conductive shoes, covers, or heel straps are required, as well as conductive casters on chairs or carts, and conductive seat covers.

Factors in Selecting Conductive Floors

Because conductive floors are performance-oriented, and this performance is critical to end-use applications, there are several factors to be considered in their selection. Is the conductive floor listed by Underwriters Laboratories? The UL listing indicates periodic testing and conformance with the electrical resistance requirements of UL 779, "Standard for Electrically Conductive Floorings." Does the floor meet Federal Specification SS-T-312B for vinyl tile? This specification

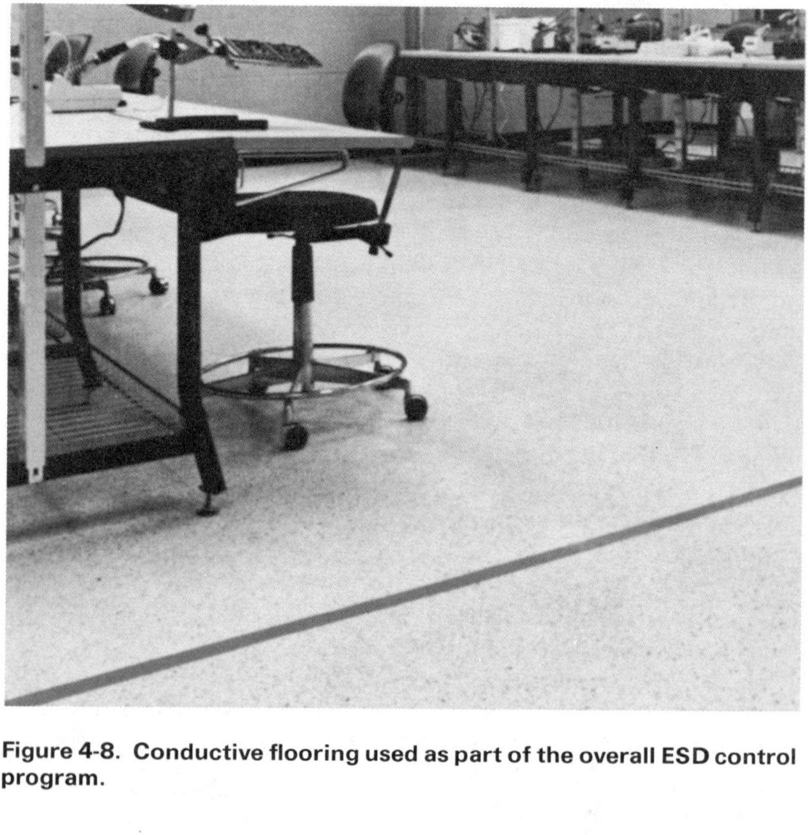

Figure 4-8. Conductive flooring used as part of the overall ESD control program.

Figure 4-9. Measuring the electrical resistance of conductive flooring with an ohmmeter.

covers basic physical properties of the material such as dimensional stability, dimensional tolerances, and flexibility. Materials meeting SS-T-312B for vinyl tile install readily, have a tight-fitting, attractive finished appearance, and conform to floor irregularities without cracking. (See Figures 4-8 and 4-9.)

Installation and Maintenance

Vinyl conductive floors can be installed over most subfloors, although some subfloor conditions may require special attention. It is advisable to check with the manufacturer or flooring contractor prior to making final commitments. Actual installation of a conductive floor by a competent flooring contractor is good insurance in avoiding installation problems.

Maintenance of conductive floors requires no special procedures or equipment. In fact, they are maintained with normal cleaning and dry buffing. The use of wax or a floor finish can build up an insulating layer of film on the floor and increase the floor's resistance, thereby reducing its effectiveness. The simple precaution is: *do not wax.*

Conductive Footwear

If you implement a floor program and ignore the question of footwear, you don't have a program at all. Appropriate footwear is essential with all of the alternatives under discussion. Leather-sole shoes generally work well with conductive floor surfaces. Some composition soles also work. Insulating soles such as crepe or thick rubber are not adequate. The general rule is to use conductive shoes, conductive foot or heel straps, or leather soles.

It is said that you can't tell employees what to wear. Yet in thousands of industries employees are required to wear safety shoes and safety glasses. And in the electronics industry we require wrist straps or special cotton clothing. Enforcing footwear standards is not easy, but it is basic to the issue of control.

CHOOSING THE RIGHT PROGRAM

Choosing the right floor management program is not as difficult as it might seem, but it does require a careful analysis of static control needs and how to meet those needs. This analysis starts with defining the extent of the facility's static problem. Identify the parts, assemblies, or products that are static-sensitive and classify them according to sensitivity. Quantify losses due to static when you prepare the cost/benefit analysis that will win upper management support for the program. Determine the physical areas where static damage is likely to occur.

Although every facility is unique, there are five common factors to consider in developing a floor management program:

1. The sensitivity of the components in use.
2. The physical areas to be protected.
3. The mobility of personnel and equipment.
4. The potential for management control.
5. The cost/benefit analysis.

Component Sensitivity

Component sensitivity is the first consideration for selecting a floor management program—not only the sensitivity of the components in use today, but also the increased sensitivity of the components likely to be in use in the future.

In areas handling the most sensitive components, conductive flooring or conductive mats provide greater static protection than the other alternatives. In areas using less sensitive components, antistatic mats, antistatic flooring, or topical treatments may provide the required protection. Antistatic carpeting may be suitable in less sensitive nonmanufacturing environments.

Mobility

The greater the mobility of personnel and equipment, the greater the need to provide environmental control beyond the localized protection afforded by wrist straps or floor mats. In manufacturing areas where wrist straps are inconvenient or mats are too localized, conductive flooring, topical treatments, or the combination of topical treatments and floor mats can provide protection for personnel who move about.

Table 4-2. Floor Management Alternatives: Suggested Use Guidelines.

	FLOOR MATS				TOPICAL TREATMENTS			FLOOR COVERINGS		
	CC	CL	RA	RL	CPC	TA	CAFF	CVF	AVF	AC
COMPONENT SENSITIVITY										
Least sensitive			X			X	X		X	X
Moderately sensitive	X	X	X	X	X	X	X	X		X
Highly sensitive	X	X		X	X			X		
PHYSICAL PLANT										
Existing facility/area	X	X	X	X	X	X	X			
Remodeling								X	X	X
Additons/new buildings								X	X	X
Clean rooms		X	X	X		X	X	X	X	
MOBILITY										
Least mobile	X	X	X	X						
Moderately mobile	X	X	X	X	X	X	X	X	X	X
Highly mobile					X	X	X	X	X	X
POTENTIAL FOR MANAGEMENT CONTROL										
Least potential						X	X			
Moderate potential	X	X	X	X	X					
High potential								X	X	X

LEGEND

CC: Conductive, Carbon-Filled
CL: Cushioned Laminate
RA: Rigid Antistatic
RL: Rigid Laminate
CPC: Conductive Paints/Coatings
TA: Topical Antistat
CAFF: Conductive/Antistatic Floor Finishes
CVF: Conductive Vinyl Flooring
AVF: Antistatic Vinyl Flooring
AC: Antistatic Carpeting

Some brands may not qualify for some clean rooms.
Carbon-filled sheet goods may not qualify for some clean rooms.

TEN TIPS FOR ESD CONTROL

- Store and transport MOSFETs in closed, conductive containers.
- Remove MOSFETs from containers only after grounding at a static-controlled work station.
- Wear static-dissipative outer clothing when handling power MOSFETs and be grounded at all times.
- Use grounded static-dissipative floor coverings, or treat floors with a static-dissipative finish.
- Use tables that have grounded static-dissipative coverings, or treat tabletops with a static-dissipative finish.
- Avoid insulating materials of all kinds.
- Use antistatic materials in one-time applications only.
- Always use a grounded soldering iron to install MOSFETs.
- Test MOFETs only at static-controlled work stations.
- Use these protective measures simultaneously, and be sure all personnel handling MOSFETs are trained in ESD control.

REFERENCES

1. W. J. Kirk, Jr., L. S. Carter, M. L. Waddell, "Control of Electrostatic Damage to Solid State Devices," *Technological Spinoff Report,* No. BDX-613-1080, March 1974, Bendix Corp. See also "Eliminate Static Damage to Circuits," *Electronic Design,* March 29, 1976.
2. Charles Briggs, Jr., "Electrostatic Conductivity Characteristics of Workbench-Top Surface Materials," *Proceedings of 1979 Electrical Overstress/Electrostatic Discharge Symposium,* Denver, September 22-25, 1979.
3. Kirk, *op. cit.*
4. Don Yenni, "Basic Electrical Considerations in the Design of a Static-Safe Work Environment," 1979 Nepcon/West Conference, Anaheim, Calif.
5. Briggs, *op. cit.*
6. Gordon T. Baumgartner, and Michael T. Brandt, "Conductive Flooring and How it Works," *Evaluation Engineering,* May/June 1980.
7. George R. Berbeco, "Characteristics of ESD Safe Requirements for Floor Surfaces," *Proceedings of the 1982 EOS/ESD Symposium.*
8. DOD-HDBK-263, *Electrostatic Discharge Control Handbook for Protection of Electrical and Electronic Parts, Assemblies, and Equipment,* U.S. Department of Defense, May 2, 1980.
9. DOD-STD-1686, *Electrostatic Discharge Control Program for Protection of Electrical and Electronic Parts, Assemblies, and Equipment,* U.S. Department of Defense, May 2, 1980.
10. Melvin Henry Downing, "ESD Control Implementation and Cost Avoidance Analysis," *Proceedings of the 1983 EOS/ESD Symposium.*

11. "ESD Protective Material and Equipment: A Critical Review," Reliability Analysis Center, Spring 1982.
12. Stephen A. Halperin, "Static Control Using Topical Antistats," *Proceedings of the 1979 EOS/ESD Symposium.*
13. W. J. Kirk, Jr., J. Whitson, Lane S. Carter, and Max L. Wadell, "Eliminate Static Damage to Circuits," *Electronic Design,* March 29, 1976.
14. William G. Klein, "Static Electricity: Problems and Solutions in Computer Facilities," United Technical Products, Inc.
15. J. E. Lingousky, and V. E. Holt, "Analysis of Electrostatic Charge Propensity of Floor Finishes," *Proceedings of the 1983 EOS/ESD Symposium.*
16. C. Fred Mykkanen, "Component Comments," Honeywell Avionics and Defense Systems, Issue A, March 27, 1980.
17. Leslie P. Phillips, "Basic Specification for ESD Protection in Industry, *Proceedings of the 1982 EOS/ESD Symposium.*
18. C. J. Strand, A. Tweet, and M. E. Weight, "An Effective Electro-Static Discharge Protection Program, *Proceedings of the 1982 EOS/ESD Symposium.*
19. Yenni, *op. cit.*

5. THE OPERATOR'S STATION

Table 5-1 (DOD-HDBK-263) shows electrostatic voltages that can be generated by personnel in a manufacturing facility. Note that there are two columns of voltages for the same condition. The decrease in generated voltage is directly related to the increase in relative humidity.

THE CONTROLLED WORK STATION

All elements have a specific relative electrical potential. When metals or elements of different electrical potential, such as steel and copper or aluminum and carbon, are in intimate contact in the presence of moisture (the electrolyte), a low-energy electric current flows from the metal having the highest potential (the cathode). Corrosion always occurs in the anode area, while the cathode area is protected.

A static-safe work area is capable of controlling static charge on conductive objects, nonconductive objects, and people. Any work

Table 5-1. Electrostatic Voltages Generated by Personnel.

MEANS OF STATIC GENERATION	ELECTROSTATIC VOLTAGES	
	10 TO 20 PERCENT RELATIVE HUMIDITY	65 TO 90 PERCENT RELATIVE HUMIDITY
Walking across carpet	35,000	1,500
Walking over vinyl floor	12,000	250
Worker at bench	6,000	100
Vinyl envelopes for work instructions	7,000	600
Common poly bag pickup from bench	20,000	1,200
Work chair padded with polyurethane foam	18,000	1,500

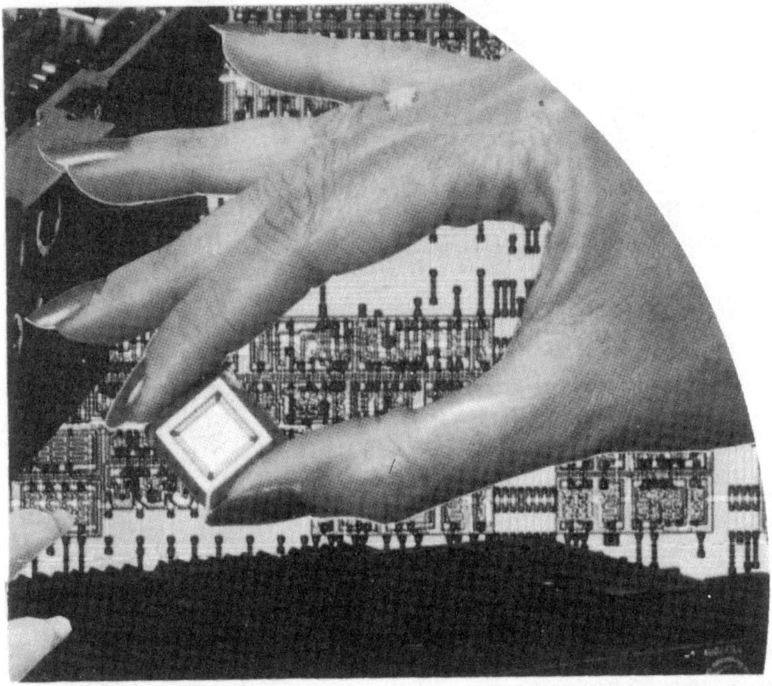

Figure 5-1. The primary source of ESD failures is contact by people

area is full of potential dangers to static-sensitive electronic devices (including MOSFETs, bipolar transistors, ECL devices, and others), but if the problem seems complex, the solution is relatively simple.

Personnel in a manufacturing environment commonly attain potentials of 5,000 volts or more just by walking across the floor—50 times higher than the level considered safe for delicate electronic components. (See Figure 5-1.)

Static Control Table Mat

This is a static-safe work surface that continually drains static charge from any conductive object placed on the mat, including parts bins, tote boxes, trays, or conductive bags. It assures that personnel and the conductive object are at the same zero static potential.

Static Control Wrist Strap

This integrates personnel into the grounding system and provides rapid (less than 0.1-second) static discharge time and a continuous drain of static charge.

Electrical current measured in amperes, rather than voltage, is the most important variable in establishing the potential danger of electricity. Three factors determine the severity of an electrical shock.

1. The magnitude of the current flowing through the body.
2. The path of the current through the body.
3. The duration of time that the current flows through the body.

The voltage necessary to produce a fatal result depends on the resistance of the person, the contact conditions, and the path through the body. Relatively small currents can be lethal, if the path includes vital parts of the body such as the heart or lungs.

Table 5-2 shows the effects of electrical current on the human body (Mil. Std. 454).

The purpose of resistors in a wrist strap, table mat, or floormat ground cord is to limit current flow, in the event a worker accidentally contacts a defective electrical device, such as a drill or soldering iron, while grounded. The amount of resistance must be sufficient to limit current flow and protect the worker, yet not so high as to limit the static decay time from the human body.

Table 5-2. Effects of Electrical Current on the Human Body.

CURRENT VALUES IN MILLIAMPS		EFFECTS ON A HUMAN
AC (60 HZ)	DC	
0-1	0-4	Perception
1-4	4-15	Surprise
4-21	15-80	Reflex Action
21-40	80-160	Muscular Inhibition
40-100	160-300	Respiratory Block
Over 100	Over 300	Usually Fatal

The recommended maximum limit of exposure to electrical current is five milliamps. One determines the amount of resistance needed to protect personnel as follows:

$$\frac{\text{Highest voltage source within reach of the worker}}{\text{5 milliamps (.005)}} = \text{Resistance needed to achieve a safe ground}$$

Don't take chances. Be sure resistors are installed on ground cords from wrist straps, table mats, and floor mats. Check them periodically to make sure they are doing their job. Remember: *be safe—not sorry!*

Care must be taken to prevent isolation at any point in the grounding system of a static-controlled work area. (See Figure 5-2.) Keep these points in mind:

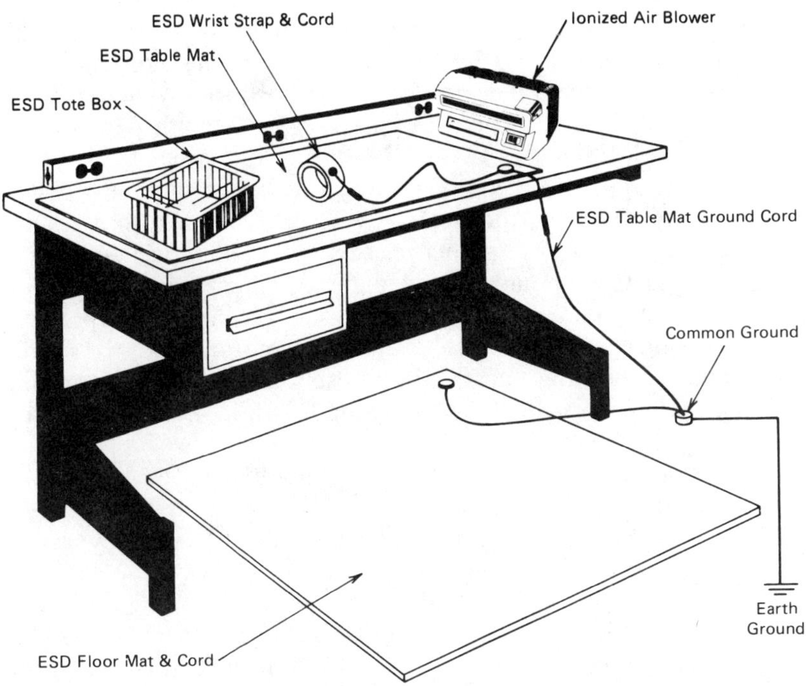

Figure 5-2. A static-controlled station.

1. A printed circuit turned over with the components facing down and laid on a foam pad creates an isolation point.The foam should be conductive.
2. All grounding should be complete to the safety ground in the facility electrical system, not earth ground.
3. Stool covers should be used to eliminate isolation between the operator and the stool. Another point of isolation is the rubber casters on the stools.

Wrist Straps

Human beings are the main source by which dangerous levels of static electricity are transmitted to sensitive electronic parts and assemblies. Since people constantly come in contact with electronic equipment during the manufacturing cycle, this has become a major concern in establishing a static control plan.

Fortunately, human skin is conductive, and static charges can be easily drained to ground by the use of a wrist strap. Therefore, since it is the most commonly used device for connecting production personnel to the grounding system, the wirst strap has become the first line of defense against ESD.

All wrist straps are made of a conductive material which, when attached firmly around the wearer's wrist and attached to a ground point through a current limiting resistor, will continuously eliminate static charges. Care should be used in the selection of wrist straps, and those having exposed metals should be avoided because of the possibility of shorting against live circuitry in work. Another potential source of problems is wearing the strap too loosely, so that it slides over the user's clothing, thus breaking the ground path and rendering itself useless.

All wrist straps must have a curent-limiting resistor in their construction. The most popular location of this resistor is close to the cuff where, if the lead wire to ground should fray, the possibility of a short bypassing the resistor will be minimized.

The accepted industry standard for the current limiting resistor in wrist straps has become one megohm, although anywhere from 250 K megohm to 10 megohm has been used. Care must be taken to ensure that the wrist strap is not connected to a table mat which has a resistance several orders of magnitude greater than the strap. This resistance will alter the decay time of the wrist strap, negating its effectiveness.

The ground connections should provide a reliable path which does not appreciably alter the resistance. In systems where the work surface provides a connection for a current limiting resistor in the grounding lead, this resistance can be added to the resistance of the wrist strap. Although this alters the decay constants, the change will not be enough to affect the strap's ability to protect.

Since wrist strap failures are not readily observable, all straps should be periodically tested to assure the integrity. A megohm resistor should be incorporated in the wrist strap for safety.

Static Control Floor Mat

This harmlessly drains to ground static charge on personnel visiting the work area. A discharge time of less than one second assures that static potential is within the safety limit before the mat is crossed and the visitor touches the components. The mat also protects one against the worker who may have forgotten to attach the wrist strap.

Ionized Air Blower

Although some companies try to bar nonconductive objects from the work area as particularly hazardous, it is almost impossible to totally eliminate such items as coffee cups, candy wrappers, and many types of employee clothing. A blower provides a constant stream of both positive and negative air ions to neutralize a static charge on nonconductors. The gentle flow of air will not disturb delicate microelectronic parts.

In the case of table and floor mats, their electrical resistances must be low enough to reduce already existing charges on conductors entering the work areas to less than 100 volts in one second. Materials having a resistance to ground of less than 1,000 megohms (10^9 ohms) will assure less than one-second discharge performance. In the case of the wrist straps, maintenance of the capacitively created voltages on an operator beneath the 100-volt safe level requires a much faster discharge time—on the order of 0.1 second. This requires a resistance of 100 megohms (10^8 ohms) to ground or less.

Because a static charge by definition cannot flow through an insulator, it cannot be removed through contact with a conductor. Grounding a nonconductor will do nothing. If the item to be discharged is an insulator, a work sheet holder, a plastic box, or even a person's clothing, ionized air is required. If, however, the charged object is a conductor such as a metal tray, a conductive bag, or a person's body, complete discharge can be accomplished by simply grounding it.

A complete static-free work station, includes a grounded conductive table top, a wrist strap, and a floor mat to discharge conductors, as well as an ionized air blower to remove the static from nonconductors.

STATIC REMOVAL FROM CONDUCTORS

In a static-free work station, static removal from conductors is accomplished by the use of grounded table tops, wrist straps, and floor mats. The rate of charge removal is restricted by the capacitance of the conductor and the resistance of the path through which it flows. An equivalent circuit for the discharge of a conductive object with capacitance (C) through a resistance (R) to ground is shown in Figure 5-3.

The voltage on the discharging body at any time t is given by this equation:

$$V = \mathrm{V_O} \exp\left[-\frac{t}{RC}\right]$$

where:

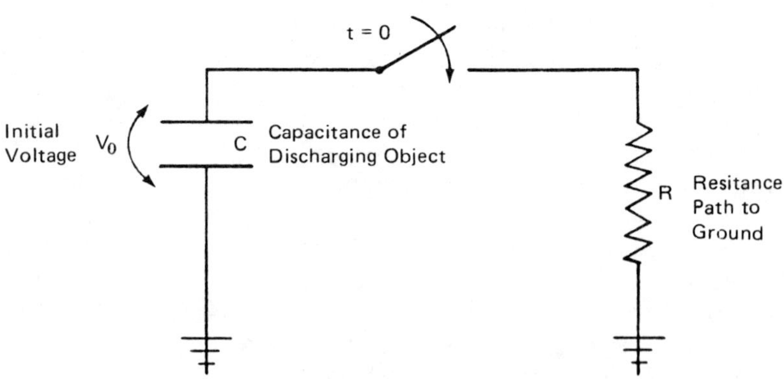

Figure 5-3. Equivalent circuit for static discharge.

V = voltage at time t (volts)
V_o = initial voltage on body (volts)
t = the elapsed time (seconds)
R = effective resistance to ground (ohms)
C = capacitance of the body being discharged (farads)

Static discharge can be thought of as the pouring of a container of water through a funnel into a sink. The sink is the equivalent of an electrical ground; the volume of the container represents the object's capacitance; the water simulates the charge (number of excess or deficient electrons); and the funnel provides a resistance to the flow of that charge. The smaller the funnel opening, the higher the resistance to flow, and the longer it takes for the water (charge) to be poured through it. By providing paths to ground capable of conducting the required electronic flow, static charge accumulations on personnel and other conductors in the work area can be effectively controlled.

In real life these paths to ground are generally made up of a number of resistances in series or parallel. For example, the resistance to ground (R) of an operator seated at a work bench and wearing a wrist strap is the sum of the wrist-strap resistance, and the resistance between the point the wrist strap is attached to the table mat and ground. if the worker's feet are resting on a conductive floor, the resistance of the path through his shoes and the mat to ground is in parallel with the wrist-strap-to-table-mat resistance.

Once the discharge time to a safe level (V) has been determined for a given set of circumstances, the theoretical discharge equation just given can be used to calculate the maximum allowable resistance (R) to ground through each of these elements of the static protective system.

Table and Floor Mats

The purpose of using a conductive floor mat is primarily to remove the static charge from transient personnel who approach the work station, such as supervisors, quality assurance personnel, and so forth. Second, it is for the worker who, after returning to the workbench, forgets to put on his or her wrist strap. The floor mat is not a substitute for good work habits, but it does provide a significant amount of passive protection around well-traveled electronics assembly areas. A person stepping onto a conductive floor mat is generally only one or two steps away from contact with static-sensitive equipment and can easily expose that device or assembly to damaging levels of static within one second. Therefore, in calcualting the maximum allowable resistance to ground (R) of the floor mat, the elapsed time (t) for discharge to under 100 volts must not exceed one second.

The conductive table mat must not only provide a surface that is free of static charge on which to work, but also remove the static charge from conductive items placed on it. For example, if a conductive tote box or bag of electronic components is carried to the table from stores, both the person carrying it and the box may be highly charged. If no floor mat is available to discharge the person and box as he approaches the table mat must do so—and do it within one second.

As mentioned earlier, in addition to the resistance of the discharge path to ground, it is the capacitance of the object being discharged that governs the discharge time. Personnel capacitance varies widely from one industry source to another. Values of 100 pF (picofarads) to 4,000 pF have been reported (200 pF being about average). Similarly, conductive tote boxes have been found to vary in capacitance from 35 to 3,000 pF, depending on their size and shape. For the purpose of calculating a maximum allowable resistance to ground for both table and floor mats, a value of 200 pF is suggested.

Although the initial voltage to be dissipated can vary, it has been well documented that personnel in a manufacturing environment commonly attain potentials of 5,000 volts or more by just walking across the floor. To be effective, therefore, both the conductive table mat and the floor mat must have a low enough resistance to ground to allow discharge from 5,000 volts down to the 100-volt safe level.

Calculation of the maximum resistance to ground for the effective dissipation of these potentials proceeds as follows:

$$V = V_O \exp\left[-\frac{t}{RC}\right]$$

where:

V = 100 volts (safe level)
V_O = 5,000 volts (initial voltage of object or person)
t = 1 second (allowable time for discharge to V)
C = 200 pF
R = maximum allowable resistance to ground in ohms

$$100 = 5000 \exp\left[-\frac{1}{200 \times 10^{-12}\,R}\right]$$

$$\ln 50 = \frac{1}{200 \times 10^{-12}\,R}$$

$$R = \frac{1}{3.9 \times 200 \times 10^{-12}} = 1.3 \times 10^{9} = 1300 \text{ megohms}$$

From these calculations, a resistance to ground through either a conductive table mat or floor mat of 10^9 ohms assures that discharge to a safe level (100 volts) will occur within one second. For all practical purposes, this minimizes the chance that a static-sensitive electronic component will be exposed to a static-caused electrical overstress.

Tests were made by the static controls systems division of 3M, using the actual static discharge of a conductive tote box placed onto various table-mat materials, and the charge removal from personnel on conductive floor mats. The resistance to ground from a conductive tote box placed onto a table is the sum of the contact resistance between the bottom of the box and the table surface, and the resistance through the table mat from the point of contact to the ground connection. Beyond this, a current-limiting safety resistor is used to protect the operator from a painful electrical shock, should he or she contact a hot electrical circuit. In this case a 1-megohm resistor is used. Figure 5-4 illustrates the experimental configuration used to make these measurements. The electrical analogue of the tote box on the grounded table is also shown.

The discharge measurements were made by charging the conductive tote box to 5,000 volts and then placing it onto a one-square-foot piece of the table mat under test. An electrometer was used to monitor the voltage on the tote box during the discharge. The results of this testing are summarized in Table 5-3 for several materials.

As predicted theoretically, materials having a resistance to ground of less than 10^9 ohms are sufficiently conductive to discharge objects in less than one second (see Table 5-4). The discharge times of materials having low resistance to ground (under 1 meghom) were extended to one millisecond by the current-limiting safety resistor in the ground cord.

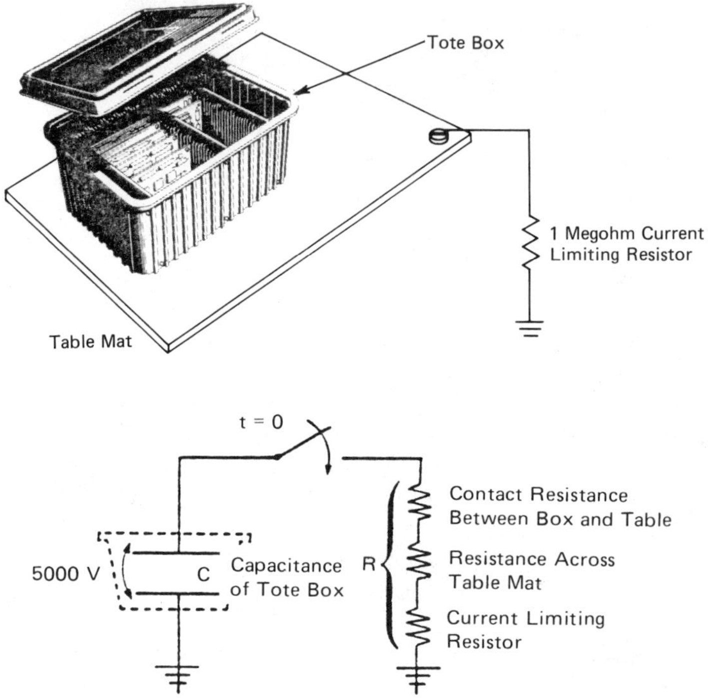

Figure 5-4. Discharge from conductive tote box to ground.

Table 5-3. Maximum Allowable Resistances and Discharge Times For Static-Safe Operations.

READING FROM OPERATOR THROUGH:	MAXIMUM TOLERABLE[a] RESISTANCE	ACCEPTABLE DISCHARGE TIME
Floor mat to ground	1000 Megohms	Less than 1 sec
Table mat to ground	1000 Megohms	Less than 1 sec
Wrist strap to ground	100 Megohms	Less than 0.1 sec

[a] Minimum resistance to ground for each is one megohm.

Table 5-4. Static Discharge Time For a Conductive Tote Box Placed Onto Various Table Mats.

TABLE MAT MATERIAL[2]	TOTAL MEASURED RESISTANCE TO GROUND (R)[3] (IN OHMS)	ACTUAL DISCHARGE TIME FOR TOTE BOX 5000 VDC TO 100 VDC (IN SECONDS)
Stainless Steel	1×10^6	.001
Carbon Impregnated EVA Copolymer	1×10^6	.001
Conductive Vinyl Laminate	1×10^8	.060
Melamine Formaldehyde	3×10^{11}	300 volts after 10 minutes
Antistatic Polyethylene	4×10^{12}	800 volts after 10 minutes
Wood	1×10^{13}	1000 volts after 10 minutes

(1) Capacitance (C) of tote box = 138 pF; size = 17 cm wide, 24 cm long, 15 cm high.
(2) All samples preconditioned for 48 hours and tested at 30% R.H.
(3) Total resistance to ground includes 1 = megohm current-limiting safety resistor.

The testing of materials for use as floor mats was conducted in the same manner (see Figure 5-5). The total resistance to ground in this case is the sum of the resistance through the person's shoes (including their contact resistance), the resistance across the floor mat to the ground connection, and the current-limiting safety resistor. Figure 5-6 illustrates the electrical analogue of this resistive network.

The 1-megohm current-limiting safety resistor alone cannot slow the discharge by more than a millisecond. Furthermore, as noted earlier, certain types of shoe soles such as crepe or heavy rubber can completely prevent the discharge of a person standing on any type of conductive floor mat.

Table 5-5 summarizes the results of a laboratory evaluation of the resistance to ground through various types of shoe soles. The data obtained are consistent with those published by the U.S. Bureau of Mines after analyzing the footwear of over 130 people in hospital operating rooms.

The resistance of the leather-soled and composition shoes can be changed significantly by such things as their age, contamination on the soles, and—probably most significantly—by foot perspiration. Leather and composition shoes have sufficient conductivity to permit adequate personnel discharge.

Crepe and thick rubber soles can be a problem; their use should be

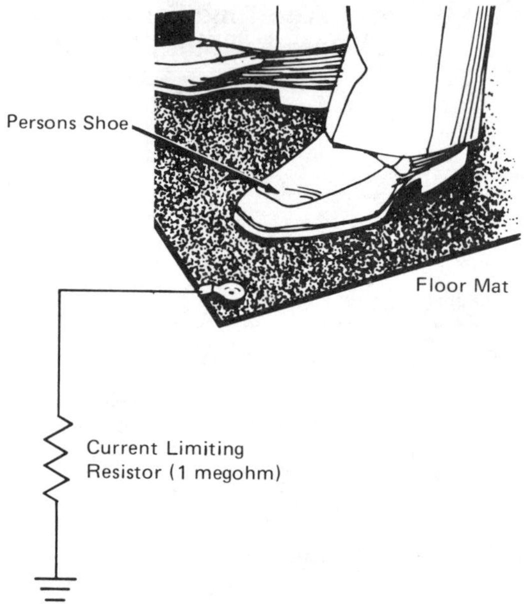

Figure 5-5. Discharge of person through shoe on coductive mat.

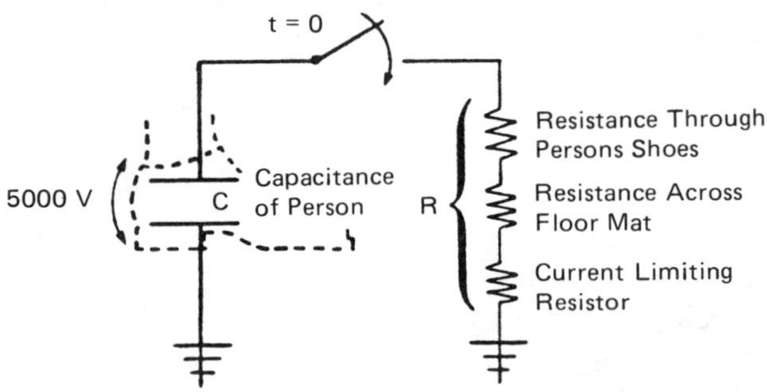

Figure 5-6. Circuit of discharge through shoe to conductive mat.

Table 5-5. Effective Shoe Resistance.

SHOE TYPE	EFFECTIVE RESISTANCE THROUGH ONE SHOE TO GROUND
Leather Soles	10 to 100 megohms
Composition Soles	175 to 750 megohms
Crepe and Thick Rubber Soles	1000 to 500,000 megohms

Table 5-6. Static Discharge Time For a Person Wearing Leather Shoes Stepping Onto Various Floor Mats.

FLOOR MAT MATERIAL(2)	TOTAL MEASURED RESISTANCE TO GROUND (IN OHMS) (3)	DISCHARGE TIME FOR PERSON FROM 5000 VDC TO 100 VDC (IN SECONDS)
Stainless Steel	4×10^7	.010
Carbon Impregnated EVA Copolymer	4×10^7	.010
Conductive Vinyl Laminate	1×10^8	.080
Melamine Formaldehyde	3×10^{11}	200 volts after 10 minutes
Antistatic Polyethylene	4×10^{12}	700 volts after 10 minutes
Wood	1×10^{13}	1000 volts after 10 minutes

(1) Capacitance (C) of person = 196 pF; shoe resistance = 39 megohms.
(2) All samples preconditioned for 48 hours and tested at 30% R.H.
(3) Total resistance to ground includes shoe resistance (39 megohms) and current limiting safety resistor (1 megohm).

discouraged in areas where static-sensitive devices are being handled. A more practical solution to this problem is to require the use of commercially available conductive heel straps, safety chains, or even booties in areas where highly static-sensitive devices are handled. (See Figure 5-7.)

For complete safety as well as high performance in discharging static accumulations, materials having a resistance to ground of between 1 and 1,000 megohms (10^6 to 10^9 ohms) should be selected for use in both conductive table and floor mat applications.

Wrist Strap

The purpose of the wrist strap is to provide a permanent path to ground for the individual operator, thus preventing unsafe voltage levels from being generated during ordinary work-related movements.

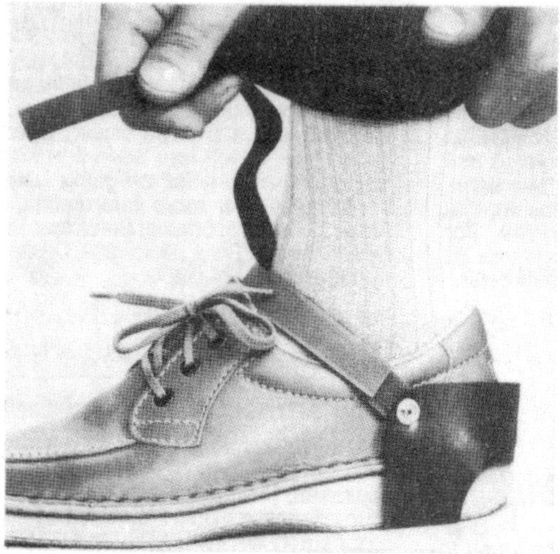

Figure 5-7. Commercially available heel strap.

At Western Electric, personnel seated at a workbench in an electronics manufacturing area were found to have residual levels averaging about 500 volts. These potentials were due not only to triboelectric or frictional charge generation (sliding around on chairs, interactions between layers of clothing, etc.), but also perhaps to another common source of voltage generation: the potential developed on personnel by changes in body capacitance.

The capacitance of a person is a function of his or her proximity to other objects. This capacitance can change significantly during simple movements such as leaning forward in a chair, raising one or both feet from the floor, or standing up.

The voltage measured on a person is in theory equal to the total charge on his body (number of excess or deficient electrons) divided by his capacitance. Even though the measured voltage on a person is zero, some induced charge will exist on his body owing to charges on his clothing, on the floor under him, and so forth. Then, if the person is ungrounded, so that the net charge on his body remains constant, changes in his cpacitance will cause measureable changes in his potential. These voltages, which are generated rapidly, may occur while the operator is already contacting a static-sensitive device. To be

Table 5-7. Capacitance Changes in Personnel in a Work Area.

DESCRIPTION OF MOVEMENT	% CHANGE IN CAPACITANCE DUE TO MOVEMENT		
	INITIAL CAPACITANCE	FINAL CAPACITANCE	% CHANGE
Person seated, raising one foot	192	163	15% Decrease
Person seated, picking up both feet and placing them on the foot rest	192	129	33% Decrease
Person seated, leaning forward in chair (desk-type chair with back)	192	184	4% Decrease
Standing person raising one foot	167	141	16% Decrease
Seated person standing up	192	167	13% Decrease

Table 5-8. Wrist Strap Resistance to Ground.

RESISTANCE TO GROUND (MEGOHMS)	PEAK VOLTAGE CREATED BY MOVEMENT (VOLTS)
0.5	<1
1.0	<1
10.0	6
100.0	85
1000.0	240
∞ (no wrist strap)	440

effective, the wrist strap must maintain the operator's potential under 100 volts during such movements. For optimum static protection, the total resistance through the wrist strap to ground should be between 1 and 100 megohms (10^6 to 10^8 ohms). (See Table 5-8.)

STATIC REMOVAL FROM NONCONDUCTORS

Tight control and restriction of nonconductors from static-sensitive work areas is the recommended solution to the problem of charged insulators. Everything from candy wrappers to work sheets, coffee cups, and even an operator's clothing can be a problem. For this reason the ionized air blower is a necessary part of the protective system. An ionized air blower neutralizes static charges on nonconductive materials by supplying them with a constant stream of both positive and negative air ions. The charged object attracts the oppositely charged ions, thereby neutralizing itself.

Prior to the establishment of any static protective system is the education of all operating personnel in the proper handling of static-sensitive devices. Once static awareness has been created, the proper equipment with which to execute these protective measures may be selected. This includes conductive table mats, floor mats, and wrist straps to discharge conductive materials, and appropriately placed ionized air blowers for discharging nonconductors. It is their combined purpose to maintain the voltage in the work area at less than 100 volts. Such equipment is a sound investment in reliability at any location where static-sensitive electronic components or assemblies are routinely handled.

Conductive Packaging

The importance of conductivity in static protection is not limited to the table mat, floor mat, and wrist strap. Whenever static-sensitive devices leave the safety of controlled environments, they must be packaged or held in containers that provide static protection. (See Figure 5-8.)

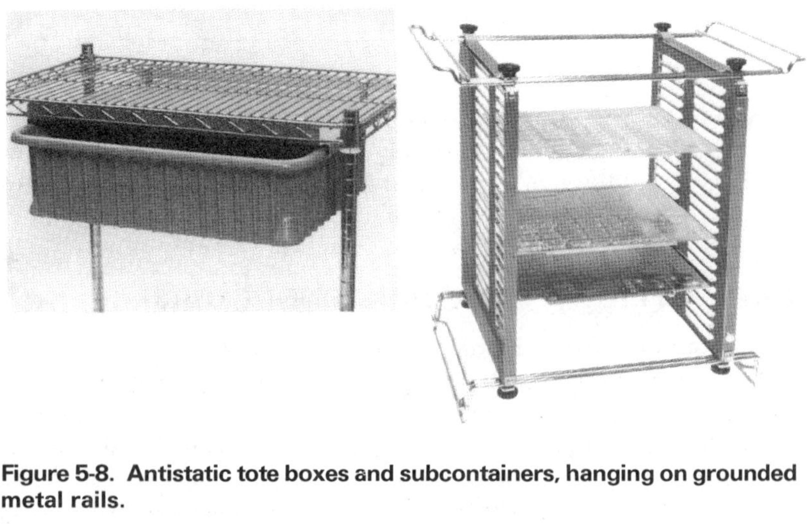

Figure 5-8. Antistatic tote boxes and subcontainers, hanging on grounded metal rails.

A good deal of attention has been given to the required electrical properties of materials for transporting and storing static-sensitive electronic components. These materials must not only remain free of triboelectrically generated charges, but must also shield the contents from the influence of external electrostatic fields.

High-resistivity materials, referred to as antistatic, are capable of preventing triboelectric charge generation. The inherent lubricity of their surfaces, coupled with a slight conductivity (up to 10^{14} ohms per square), is generally sufficient to prevent the accumulation of charges due to rubbing on the parts or on the bag itself. But the packaging problem cannot be solved by these materials alone, since they do not provide protection from the influence of external static fields.

In order to provide the necessary Faraday Cage protection (electrostatic shielding), the packaging must have at least one highly conductive surface or overwrap. Conductive tote boxes, kitting trays, and other temporary carriers used for the in-plant transfer of devices between static-safe work stations must also provide complete static protection for their contents. Metal as well as highly conductive plastic containers, having a volume resistivity of under 10^5 ohm-cm, have been used successfully in this application for many years. The surface resistivity of this shielding layer must not exceed 10^4 ohms per square.

Monitoring

Once a safe work station has been installed for the protection of static-sensitive circuitry, it must be monitored to ensure the effectiveness of that station on a regular basis. Without such examination, parts of the system may become ineffective; a wrist strap may be inadvertently disconnected; the ground cord to a table mat may become worn or cut by equipment placed on it; or dirt or contamination may eventually affect the surface conductivity.

A portable megger can be used to give a reliable measure of the resistance to ground of the work station and verify its functionality. Unlike an ordinary VOM, these devices are capable of reading resistances on the order of 10^9 ohms at potentials approaching those considered critical in acutal static discharge situations, that is, above 100 volts.

HOW TO HOOK UP A TABLE MAT

To ensure the maximum static discharge protection and protect personnel against the possible electrical shock, the ground connection of your mat should be made by a trained electrician.

The mat must *never* be connected directly to a *hard ground.* A hard ground is a connection to ground either directly or through a low impedance. A *soft ground* is mandatory. This is a connection to ground through an impedance (resistor) sufficiently high to limit current flow to safe levels for personnel (normally, 5 milliamperes). Impedance needed for a soft ground is dependent upon voltage levels which could be contacted by personnel near the ground.

Under normal circumstances where 110-volt power sources are within reach of a worker, a one-megohm) resistor or equivalent must be connected in series with your mat grounding cord. If higher voltage sources are within reach of the worker, then the resistance needed to achieve a soft ground must be calculated. The following formula may be used:

$$\frac{\text{Highest voltage within reach}}{\text{5 milliamps (.005)}} = \text{Impedance (resistor) needed to achieve soft ground}$$

Based on this formula, a one-megohm resistor provides an ample safety margin, even if 240-volt power sources are within reach.

Caution must be observed in employing parallel paths to ground that could reduce equivalent resistance of persons to ground to unsafe levels. Personnel movements in conjunction with wrist straps, table mats, and conductive floors or floor mats could result in such parallel paths.

Figure 5-9 shows a parallel wiring diagram, while Figure 5-10 shows a series diagram. The series diagram may be used, so long as the cumulative resistance of the wrist strap resistor and the table mat resistor do not exceed 10 megohms. In no case should the table mat and a wrist strap be connected as shown in Figure 5-11, with wrist strap, mat, and ground cord all in series. Mats equipped with studs at both sides are intended for left-hand or right-hand applications, for the worker's convenience, not for making series connections of the three components.

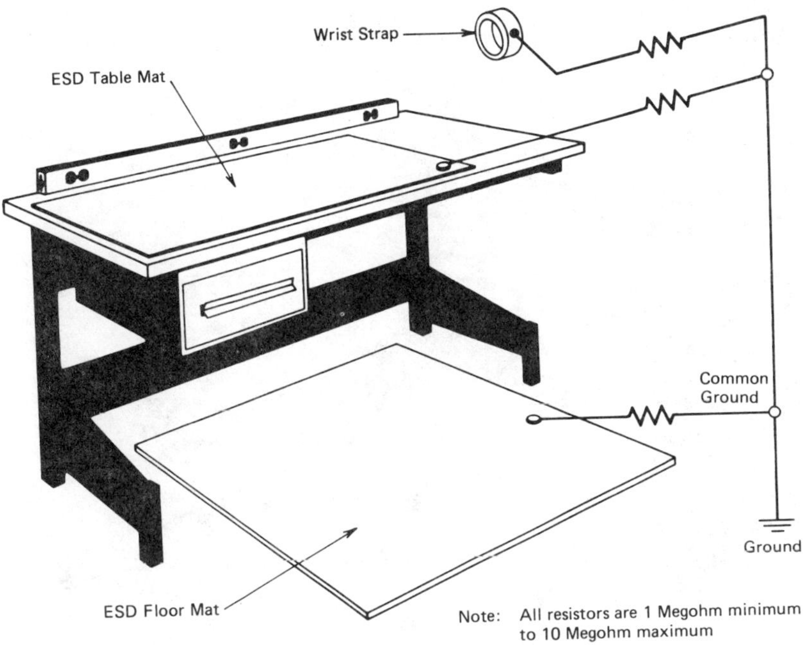

Figure 5-9. Parallel circuit hookup.

Periodic checks should be made to make sure that an electrically conductive leakage path, limited by the proper resistance, exists between each of the ESD devices to ground. All ground cords should be examined for wear, and every resistor should be checked on a regular basis to make sure it is functioning properly.

WRIST STRAPS

Human skin is the main vehicle by which dangerous static levels can be transmitted to sensitive circuits and devices. Fortunately, human skin is conductive, making it possible to drain static charges to ground. The wrist strap which connects its wearer to ground has become one of the most commonly used products in the fight to control ESD.

All wrist straps use a conductive material inside a cuff which is strapped around the wearer's wrist to make electrical contact with the

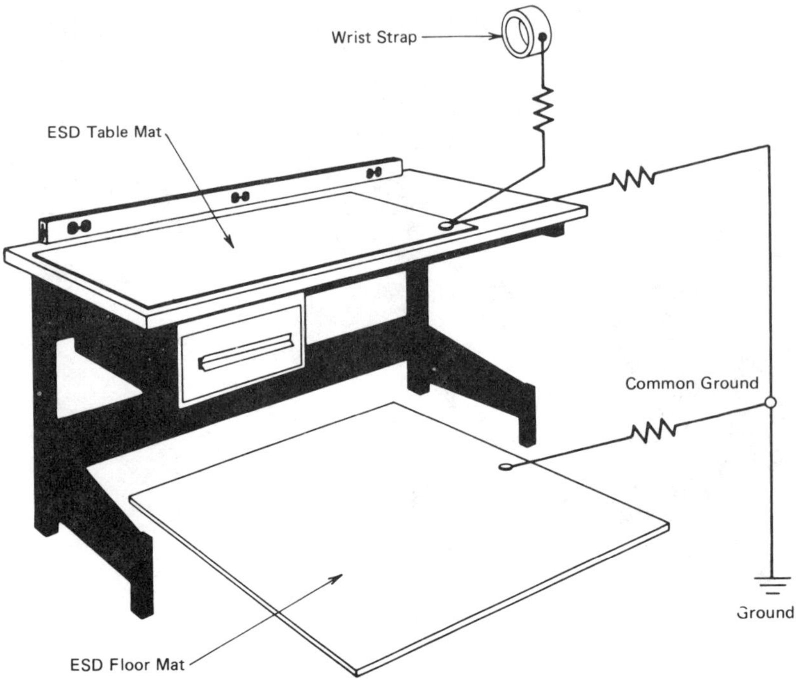

Figure 5-10. Series circuit hookup.

skin. This cuff is connected through a current-limiting resistor and lead to a ground point. (See Figure 5-12.) Available wrist strap constructions include metal wire beads, conductive plastics, Velcro bands with metal straps, "Speidel"-like watch bands, and cloth bands with wires woven into the fabric so they contact the skin but are not exposed on the outside. (See Figure 5-13.)

When selecting a wrist strap, consider exposed metal surfaces which could accidentally short-out live circuitry that the wearer is working on. Designs which incorporate wire beads and conductive plastics are reportedly susceptible to deterioration and corrosion because of body oils. A further shortcoming frequently cited is that the cuff can be too loose, resulting in intermittent contact with the wearer's skin. A loose cuff may slip over the clothing of the user, breaking the ground path

*How well does the lead handle continuous flexures

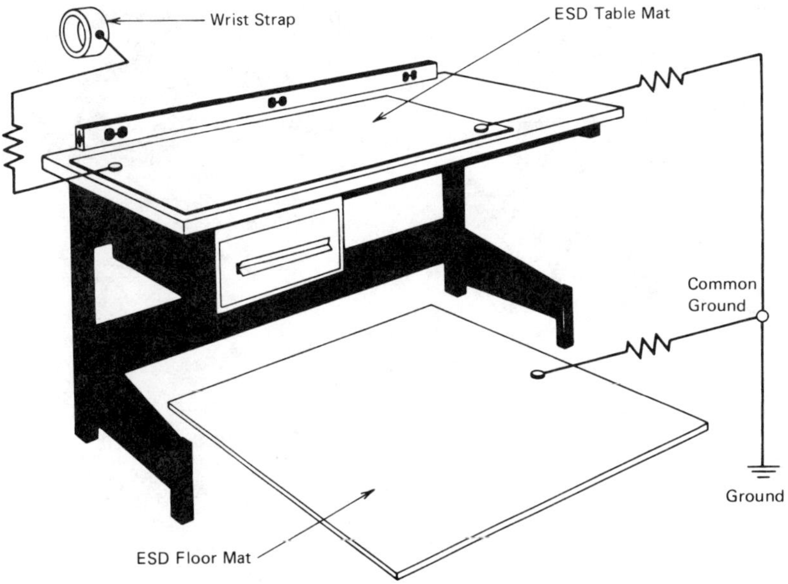

Figure 5-11. Wrist strap, mat, and ground cord all in series. *Not Recommended.*

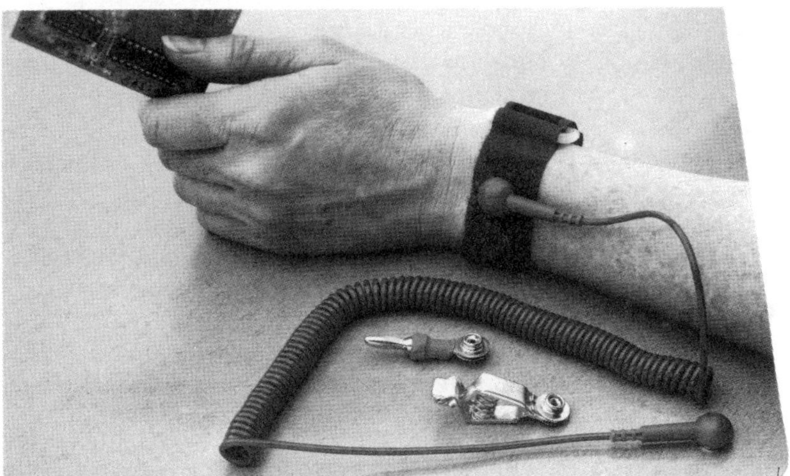

Figure 5-12. A wrist strap.

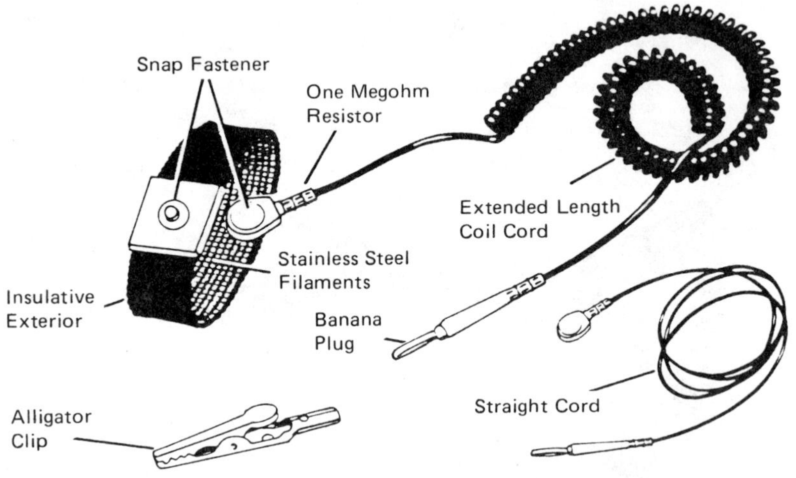

Figure 5-13.

permanently. Also, discomfort leads some users to wear the cuff looser than was intended.

All wrist straps should incorporate a current-limiting resistor in their construction. Some wrist straps have the resistor in the ground end, while others place it at the cuff end of the lead. The cuff location is preferred, owing to concerns that the lead wire insulation might become frayed, permitting a short which would bypass the limiting resistor.

A one-megohm resistor is becoming an industrial standard. This value has appeal, because the decay time is short enough for most circumstances and spares the operator from sensing the static discharge.

The wrist strap user must clearly understand how the wrist strap is used and how its use relates to the materials to which it might become connected. A common error is to connect a wrist strap to an antistatic tabletop mat which has a resistivity several orders of magnitude greater than the limiting resistor. This resistance alters the decay-time constant, negating the effectiveness of the wrist strap.

The wrist strap selected must ensure 100% contact with the operator's skin. The Velcro-type wrist strap that allows the operator to select the tension can easily be set so loose that the only contact with the skin is

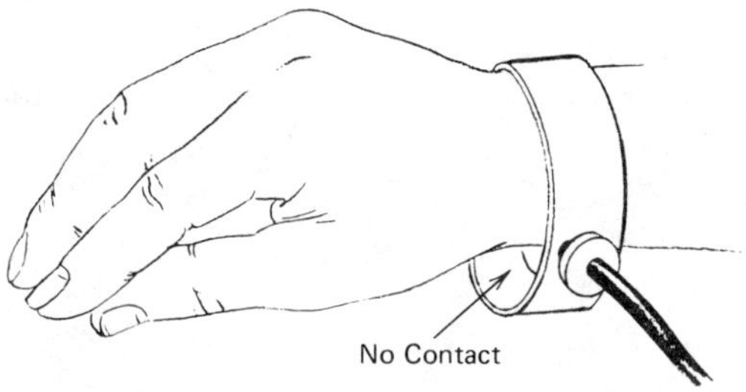

Figure 5-14. An improperly tightened wrist strap, providing insufficient contact.

that resulting from the pull of gravity. Thus the strap may be in contact only at the top of the wrist. (See Figure 5-14.)

Clothing

Zaps normally occur when an ESD-sensitive component comes into contact with a material with static buildup. However, a zap can occur without direct contact. The static-laden material, item, or person generates an electrostatic field which can zap a component by induction. In a normal electronic testing or assembly operation, induction is especially important when considering static buildup on apparel. The component does not even have to come into contact with the person's garment to be zapped!

To develop a static-free work environment, a number of products have been designed to prevent static buildup by safely grounding or neutralizing the electrons. The major products used are:

- Conductive flooring
- Conductive chair
- Conductive table mat
- Conductive wrist strap

What's missing here is the ability to drain the static buildup from the person's clothing. The answer: static-dissipative garments. (See Figure 5-15.)

Figure 5-15. Static-dissipative clothing.

Apparel made from wool, silk, and synthetic fibers generates static buildup. In a normal work situation this can be as much as 1,500 volts. Heavy activity will cause much larger amounts. Even cotton, especially in low humidity, can cause static buildup. All those static-free work station products together will not effectively knock out the static built up in normal apparel on a person in typical electronic testing and assembly work activities.

For a number of years, firms have marketed garments with a 65% polyester/34% cotton/1% steel plain weave. These fabrics and garments were developed for use in hospital operating-room apparel. There is a static concern in operating rooms, since if certain anesthetics are used, explosions can occur. This is an excellent product for the operating room, but it is not good enough for the static sensitivities of today's electronic components. A new fabric developed for static-dissipative garments is a 50% polyester/48% cotton/2% steel which is mercerized, super precured, preshrunk 2%. The individual needing these garments will be handling chips, microcircuits, semiconductors, resistors, and other types of electronic components during assembly, testing, packaging, cleaning, or repairing.

There is also a growing need for operators in computer rooms to be in static-free work environments. Other needs are just now becoming known. With the continuing miniaturization and sophistication of electronic chip technology, we will see greater ESD sensitivity of electronic components. Today's 64K chip for computers is evolving into tomorrow's 256K chip. This will increase the capabilites of tomorrow's computers beyond our comprehension. It will also increase the ESD sensitivity and the need for static-dissipative garments.

REFERENCES

1. L. A. Schrier, "Electrostatic Damage Susceptibility of Semiconductor Devices," 1978 International Reliability Physics Symposium.
2. O. J. McAteer, "Electrostatic Damage in Hybrid Assemblies," *Proceedings, 1978 Annual Reliability and Maintainability Symposium.*
3. W. J. Kirk, Jr., L. S. Carter, M. L. Waddell, Control of Electrostatic Damage to Solid State Devices," *Technological Spinoff Report,* No. BDX-613-1080, March 1974, Bendix Corp. (RAC Acct. No. 11704). See also *Electronic Design,* March 29, 1976.
4. *Proceedings of the 1977 Reliability Analysis Center Electrostatic Discharge Seminar,* August 29-30, 1977.
5. J. T. Petrick, "Discharge of an Electrostatically Charged Human," *Proceedings of the 6th Symposium on Electroexplosive Devices,* San Francisco, July 8-10, 1969.
6. T. S. Speakman, "A Model for the Failure of Bipolar Silicon Integrated Circuits Subjected to Electrostatic Discharge," *12th Annual Proceedings Reliability Physics,* 1974.
7. F. L. Jones, *The Physics of Electrical Contacts,* Oxford University Press, 1957.
8. P. G. Guest, V. W. Sikora, B. Lewis, "Static Electricity in Hospital Operating Suites," U.S. Bureau of Mines, RI 4833, January 1952.
9. J. R. Huntsman and D. M. Yenni, "A New Technology in Transparent Static Protective Bags," *Evaluation Engineering,* July/August 1978, pp. 48-50, and *Quality,* December 1978, pp. 18-21.
10. Don Yenni, "Basic Electrical Considerations in the Design of a Static-Safe Work Environment," Static Control Systems/3M, St. Paul, Minnesota.

6. MATERIALS FOR USE IN ESD ENVIRONMENTS

MATERIALS FOR STATIC CONTROL

Since static electricity can and very likely will damage solid-state devices, this chapter will discuss packaging materials that can be used to provide protection for these devices. The intent is to provide specifications which will be useful in the procurement of these items, once a particular packaging method or procedure has been chosen and identified.

Static charges on personnel and their clothing, and on common plastics used in the electronics industry, are a major source of undesirable electrostatic fields. The resulting need to implement protective measures in manufacturing and transportation, higher-level product assembly, and field service has led to the development of a variety of products to meet the protective requirements of each activity.

This chapter treats several different categories of electrostatic protective-packaging materials. These categories reflect the electrostatic properties of the materials to charge triboelectrically, to dissipate a charge on a surface, or to act as a shield against electrostatic fields. The form or shape in which these materials are used is unrestricted and may embody the characteristics of more than one of the above categories.

Any packaging material must maintain its antistatic, conductive, or shielding qualities during storage, shipment, distribution, and application, being used to the point of either reuse or disposal at the user's option. As regards fundamental static-protective materials, the flexible and rigid container forms currently available are basic bags; pouches and bags that integrate static protection and mechanical cushioning; rigid DIP tubes; tote boxes; kitting containers, bins, and shipping cartons; as well as a number of special-purpose containers.

FUNDAMENTAL MATERIALS

Static-protective materials surrounding a sensitive part function in various degrees, depending on the properties of the material, to minimize charge generation; to dissipate charge by redistributing it on a surface or by providing a path to ground; and to shield the part from external electrostatic fields. Historically, following the Department of Defense's lead, materials have been categorized in order of increasing conductivity as antistatic, static-dissipative, and conductive. Currently, however, an Electronic Industries Association interim standard categorizes materials as antistatic, conductive, and electrostatic-shielding. Note that "antistatic materials" refers to a class with specific conductivity, and is not a generic description of all static-protective materials.

Safety

1. As a safety feature, a one-megohm resistor has been installed in wrist strap cord and ground cord. This resistor limits current, should an operator accidentally come in contact with a 120-volt power source. *Do not remove the molded caps from any ground cord. If they are damaged, replace them immediately.*
2. When positioning the conductive work surface, take care that it does not touch grounded conductive objects. This could create an alternate path to ground, bypassing the protective resistor.

Cleaning

Cleaning should be accomplished with mild detergent and water. If a stronger solution is needed, isopropyl alcohol may be used. Most solvents are not recommended. The work surface should be cleaned, because excessive contamination can increase contact resistance.

Temperature

The conductive film work surface should not be used over 150° F. If exposed to severe cold (40° F and colder), acclimate to room temperature before unrolling.

Electrical Continuity

Important Notice: To measure the resistance of the mat, a high-voltage ohmmeter (Megger) is required. If one uses a 5-pound, 2.5-inch diameter electrode, the resistance to ground should be greater than 10^6 ohms and less than 10^9 ohms. If the resistance is less than 10^6 ohms, it indicates an alternate path to ground (bypassing the one-megohm resistor) which should be traced down and corrected. If the resistance is greater than 10^9 ohms, it means that the ground connection is interrupted and must be repaired.

MILITARY SPECIFICATIONS AND STANDARDS

In products that protect sensitive electronic devices from static discharges, test methods and standards that must be met have their origin in a military specification or standard. A "specification" and a "standard" are different documents: a specification is a demand for a specific material or a product, while a standard shows how the material or product is to be tested, packed and labeled.

The electrical and physical properties of a material are found in the material specification, along with all the associated standards that must be met. Unless the basic material is the product itself, there is usually a second set of specifications and standards that define the product.

A black conductive bag will illustrate the point. The material from which the bag is made is covered by Mil. Spec. MIL-P-82646, entitled "Plastic Film, Conductive, Heat Sealable, Flexible." If the product was a plain roll of this film, this specification would be both the material and product specification. A bag from this material would remain MIL-P-82646, while a second specification—MIL-B-82647, entitled "Bag, Conductive, Heat Sealable, Flexible," would govern the bag manufacture.

The case of a pink poly bag or a metallized, conductive bag is identical. The material specification for both these bags is found in Mil. Spec. MIL-B-81705, entitled "Barrier Materials, Flexible, Electrostatic Free, Heat Sealable," while the product specifications for the bags themselves are found in MIL-B-117, entitled "Bags, Sleeves and Tubing, Interior Packaging."

Within each of the specifications, reference is made to the various standards that must be met. There are standards for measuring static decay (Fed. Std. 101, Method 4046) and surface resistivity (ASTM-D-257). There are standards for colors (Fed. Std. 595), and for tests made on physical properties of plastics such as tensile strength, elongation, and heat-seal strength (Fed. Std. 406), to name a few. Because of the maze of specifications and standards involved, very few manufacturers comply with all applicable specifications and standards. Be careful of what you read, since product information sheets are carefully worded.

Here is a classic example taken from sales literature: "Meets all ESD prevention requirements MIL-B-81705 Type I." *Read carefully!* Published claims may be accurate as far as they go, but "meets or exceeds static decay test of MIL-B-81705" does not say "meets or exceeds *all* requirements of MIL-B-81705."

Materials in the manufacturing environment can either help or hinder static control. All the materials in manufacturing areas will fall into four clearly defined categories of surface resistivity: insulating materials with a surface resistivity of 10^{14} ohms per square; antistatic, 10^9 to 10^{14} ohms per square; static-dissipative (shielding), 10^5 to 10^9; and conductive, less than 10^5.

For ideal protection, all conductive materials including personnel should be grounded. Since in most cases this is not practical, other protective materials and methods must be selected to fit each situation. For instance, insulating materials tend to store static electricity and have difficulty in discharging these charges. It is imperative to keep these materials away from static-sensitive components, and out of the manufacturing area entirely if possible. These materials include polyethylene (plastic bags, polystyrene styrofoam cups, and packing materials for foods), mylar, hard rubber, vinyl, mica, ceramics, plastics, and some organic materials. Only materials treated with antistatic compounds impregnated with a conductive material should be used in a static-controlled area.

"Pink poly" is often used to describe polyethylene containing an antistatic additive. Care must be taken to ensure that counterfeit pink poly is not used for static-control applications. Pink poly should be tested and/or certified before use in any controlled areas.

Conductive Materials

These are materials which are either metal or impregnated with metal, carbon particles, or other conductive materials, or whose surface has been treated with such through a process of lacquering, plating, metallizing, or printing. These materials seldom charge when separated from one another, but can produce charging when separated from nonconductive surfaces.

Antistatic Materials

When a material is antistatic, you cannot generate any static electricity on it, unless you isolate the material and charge it by induction. This is also true of any conductive material. There is a controversy between antistatic and conductive materials. Antistatic materials are necessary in electronic manufacturing, though they must be used properly. You would not consider using a carbon-loaded conductive plastic in a clean room because of its contamination, but there is no such risk with an antistatic material.

Antistatic materials are those materials that resist triboelectric charging and produce minimal static charges when separated from themselves or other materials. They consist of two basic types: those materials impregnated with migratory antistats, and those which have been treated by spraying, dipping, printing, or wiping with a topical antistatic agent to render them surface-conductive.

Electrostatic-Shielding Materials

These are materials which are capable of attenuating an electrostatic field, so that its effects do not reach the stored or contained item and produce damage.

Conductive Surfaces

Typical conductive surfaces are metal, conductive carbon, or the human sweat layer. Except in special circumstances, the separation of two conductors will not produce a net charge on either surface, because the charge will equalize through the areas still in contact. If a conductor separates from an insulator, both can retain the static charge (e.g., a person walking across a nylon carpet).

Nonconductive Surfaces

These are virtually all dry nonmetallic textiles, plastics, paper, cotton, wood, and some oxide-coated metals. These surfaces can produce charges when separated from each other or from conductive surfaces.

Antistatic Surfaces

Antistatic surfaces have a thin lubricious layer created by an antistatic agent which reduces frictional effects during rubbing and separation. Most antistatic agents are hygroscopic and create a surface layer which is both lubricious and slightly conductive. The antistatic agent may be applied topically by dipping or similar means, or be impregnated in the material so that it migrates to the material surface. Handling, wear, and environmental factors can reduce antistatic characteristics so that retreatment may be necessary.

Surface Resistivity

Surface resistivity of antistatic plastics shall be less than 1.0×10^{14} ohms per square, but greater than 1.0×10^{10} ohms per square on both surfaces as measured in accordance with ASTM D-257 or the equivalent.

Conductive Types

Materials which are surface-conductive should have a surface resistivity of less than 1.0×10^{5} ohms per square.

Electrostatic-Shielding Types

The shielding surface should be electrically continuous. The surface resistivity should be less than 1.0×10^{9} ohms per square or, in the case of volume-conductive materials, less than 10 ohm-centimeters per mil of thickness.

WHAT DO WE MEAN BY 10^4?

When we talk about electrical resistivity, we use numbers like 10^4, 10^9, 10^{14}, or 5.5×10^7. These are mathematical expressions of very large numbers. Rather than writing the number 1,000,000,000 we simply write 10^9. If we need to talk about very small numbers, we can use the same system. 0.000,000,001 can be written 10^{-9}. Table 6-1 gives the decimal equivalents of numbers from 10^{18} down through 10^{-18}.

When quantifying the number of ohms resistance in an antistatic mat, we get into the millions of ohms, and also talk about the time it takes to damage an electronic component with a static discharge in

Table 6-1. Decimal Equavalents of Numbers From 10^{18} Through 10^{-18}.

EXPONENTIAL EXPRESSION	DECIMAL EQUIVALENT	PREFIX
10^{18}	1,000,000,000,000,000,000	exa
10^{15}	1,000,000,000,000,000	peta
10^{12}	1,000,000,000,000	tera
10^{9}	1,000,000,000	giga
10^{6}	1,000,000	mega
10^{3}	1,000	kilo
10^{2}	100	hecto
10	10	deka
10^{-1}	0.1	deci
10^{-2}	0.01	centi
10^{-3}	0.001	milli
10^{-6}	0.000,001	micro
10^{-9}	0.000,000,001	nano
10^{-12}	0.000,000,000,001	pico
10^{-15}	0.000,000,000,000,001	femto
10^{-18}	0.000,000,000,000,000,001	atto

millionths of a second. If you want to talk in the millions, simply prefix your unit of measure with the prefix word *mega.* Thus 1,000,000 (1 million) ohms become 1 megohm. When you talk about a spark taking place in 1,000,000th (one millionth) of a second, simply say 1 micro second.

TURNING INSULATORS INTO CONDUCTORS

A static electrical charge can build up on both conductive and nonconductive surfaces. The difference between them shows up when they are grounded. If you charge a steel plate while holding it with teflon tongs in very dry air, the charge on the plate will remain there for quite some time. However, the instant the steel plate touches a grounded surface, the entire charge becomes mobile and moves to ground. If you repeat this process with a styrene coffee cup, however, only the charge in the immediate area of contact between the cup and ground will drain off, leaving the remainder of the charge on the cup. In the first case, steel is a good conductor, while in the second case, styrene is considered an insulator.

To get rid of a static charge, then, two elements are necessary: a conductor and a ground. The only practical ways to remove a charge from an insulator are to bathe its surface with ionized air, cover its entire surface with a conductive layer, or blend conductive materials into it.

1. The surface of an insulator can be covered by spraying a coat of topical antistatic material on the entire surface. Products can be applied from spray containers, or the insulator can be dipped. In most cases, topical antistats work because they attract moisture from the surrounding air. This thin layer of water molecules on the surface will act as a conductor.
2. The entire surface of an insulator can be coated with a conductive material. Surface coatings such as conductive paints, or thin layers of metal that can be adhered to the surface by processes called vapor coating or sputtering, may be applied.
3. During the processing of a material like polyethylene or other polymeric materials (plastics), and antistat material can be combined with the basic raw material during the extrusion process. By doing this, the entire thickness, rather than just the surface of the sheet, will be saturated with the antistat. As the sheet ages, the antistat continues to migrate to the surface, where it attracts the same moisture layer that is attracted by the spray coating or dipping method. This is how pink poly is made.
4. An insulative material such as polyethylene can be combined with a conductive material such as carbon at the raw material state. If enough carbon is added, carbon particle to carbon particle contact throughout the entire mass is achieved, and the surface as well as the mass of the insulator is turned into a conductor. Materials made in this fashion are called homogeneous. This is the process used in the manufacture of conductive bags, shunting foams, and the conductive layer of table mat material.

In the case of topical and conductive coatings, only the coated surface becomes conductive. In the case of antistat and conductive additives to raw materials, the entire material is converted from an insulator to a conductor. A clear understanding of the difference between coated and homogeneous materials is essential to an understanding of how layered materials differ from homogeneous materials in (1) conducting charges of static electricity to ground, and (2) shielding sensitive devices from the harmful effects of other electrical forces that can damage or destroy sensitive electronic devices.

SURFACTANT MATERIALS: MANUFACTURING PROCESS

The commonest method for producing antistatic properties in synthetic materials involves adding approximately 35% of surfactant, which is a hydrophilic or water-absorbing anionic, catonic, or one of the nonionic groups on the surface of these materials. They are all dependent on the relative humidity of the surrounding environment, are nonpermanent, are not chemically bound to the plastic, and can be rubbed off, migrated, or leached out of the material. If used in a hot and dry or extremely cold environment, in many instances they will dry out.

The surfactant-filled material works by conducing the static charge along its surface moisture layer to ground. In most cases the static energy is dropped across the resistive surface and dissipates in the form of heat, which also causes evaporation, adding to the loss of moisture. This phenomena accounts for the slight increase in decay time when the surfactant material is charged and discharged several times in succession. (See Figure 6-1.)

Surfactant materials provide excellent antistatic protection when used with certain considerations in an ESD control program. This type of material can be used in clean-room applications. It attenuates a standing electrostatic field and will provide sufficient protection in almost all electrostatic-discharge or induced-field situations.

Products made from this material should be fresh, handled as little as possible, and not held in storage for long periods of time before use. The addition of additives substantially reduces the mechanical integrity of the material. Sheet material of a thickness of two mils or less cannot be guaranteed to be antistatic, as there is not sufficient volume of surface to prevent its drying out and becoming a static generator. This material is not recommended for use for long-term storage of electronic parts or assemblies.

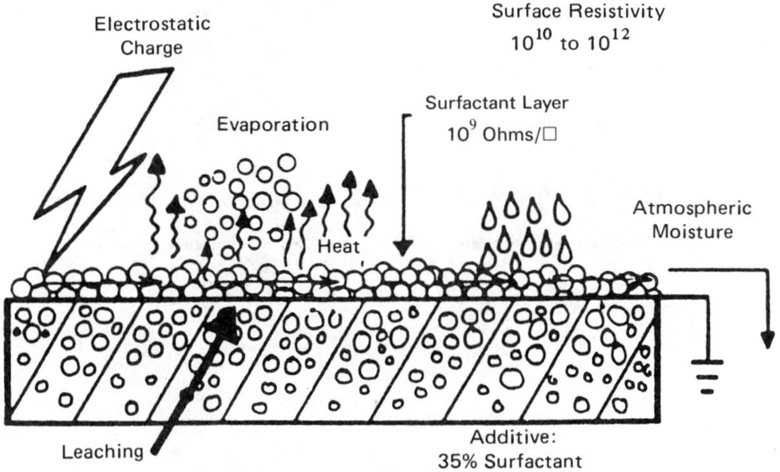

Figure 6-1. Surfactant-filled material.

LAMINATED PRODUCTS

A recent introduction to ESD control are laminated plastic products, which of course are not new to the plastics industry. These products are produced by using heat or adhesive to laminate a thin sheet of flexible plastic, such as vinyl or semirigid mica material, to a conductive layer or grid of carbon or metal, which is then laminated to either a soft plastic foam or pressed paper backing. The surface resistance varies with the manufacturer from 10^2 to 10^9 ohms per square, depending on the material used.

This material works by conducting the static charge through a small area of the volume of the surface material to the conductive layer to ground. (See Figure 6-2.) The surface is an antistatic material, while the carbon or metal layer or grid is electrically conductive. The bulk of the charge current passes through the conductive layer to ground.

Electrical current takes the path of least resistance. Threfore the amount of voltage dropped, as it passes through a small point of the volume of the material to the electrically conductive layer or grid, is dependent upon the volume resistance of the surface layer where the charge is introduced. For personnel safety, it may be prudent to place

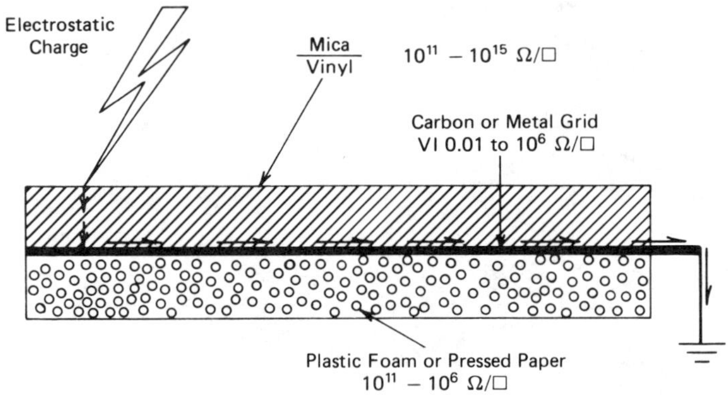

Figure 6-2. Functional surface resistivity: 10^8 to 10^{11} Ω/□.

an insulating barrier between electrically powered equipment and the laminated surface, lest the purpose of a soft ground be defeated.

LAMINATED SHEET AND BAG MATERIAL

Other forms of laminated products are the transparent metallized sheet and the insulated aluminum foil sheet material, both of which are made into bags. (See Figure 6-3). The transparent matallized bag material acts the same as the carbon impregnated plastic bag to provide a conductive surface for static charge protection. In both cases the bags (see A and B in the figure) are said to provide Faraday-cage, -screen, or -shielding protection by distributing the charge equally over the entire surface of the bag, thus protecting the sensitive parts within. By definition, however, a Faraday shield (screen, cage) is an electrostatic shield with grounded common point. Electromagnetic waves are not influenced by the shield.

Conductive carbon or metallized bags will take the full charge when a person or charged object touches them, and if insulated will hold that charge for several minutes until it dissipates the charge into the surrounding air, or until some other grounded or object touches it.

CARBON-FILLED MATERIALS

Another common method of maufacturing static protective materials is the addition of approximately 40% or even as much as 55% by volume

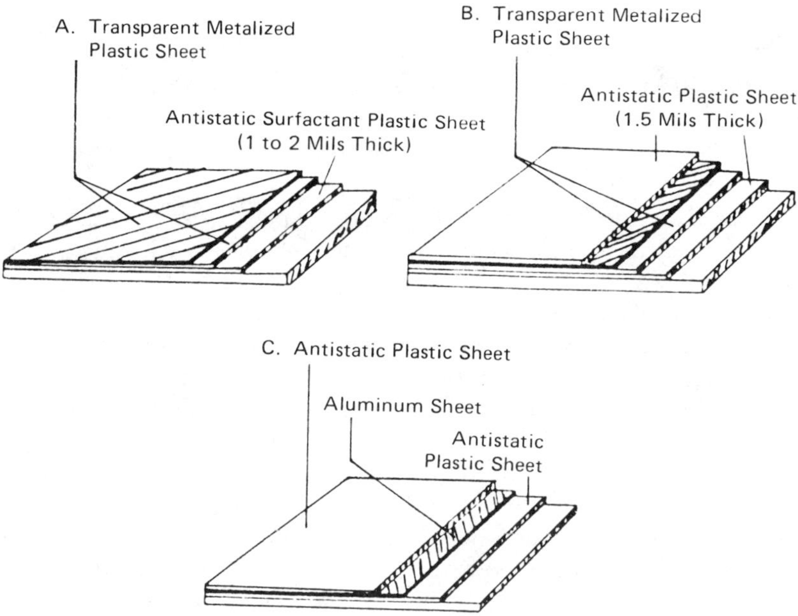

Figure 6-3. Laminated sheet and bag material.

of carbon powder into the plastic to produce an electrically conductive material. Other additives are pan carbon fiber, milled pitch carbon, aluminum flakes, and other metal flakes which produce a resistivity of from 10^2 to 10^6 ohms per square. Since the conductivity of the additive carbon powder is the easiest to control, from 10^3 to 10^6 is typical. The electrical properties of carbon powders vary considerably with particle size, degree of aggregation, and chemical purity. These variables in powdered carbon account for product variability in quality and electrical characteristics.

Most carbon-filled materials are exposed to heat as a last step in the manufacturing process, resulting in a smooth surface with a disproportionately high density of carbon which acts as a conductive electrode producing 10^3 to 10^5 surface resistivity. However, this is a conductive material mixed into an insulative plastic, and a very moderate amount of abrasion will rub away the carbon-enriched surface, leaving the insulating plastic surface.

The conductive material carries the current produced by the static charge along its enriched surface or through the volume of the material to ground. (See Figure 6-4). The resistance of conductive material is insufficient to quickly dissipate high-voltage static electric energy into heat by dissipation, as the surfactant materials do; therefore most of the energy produced by ESD must be conducted electrically through the material to ground.

Products made from carbon-impregnated plastics provide an excellent Faraday-cage effect, as long as they are grounded, but do not protect ESD-sensitive devices or assemblies from RFI or EMI. The high volume of carbon used in producing this material causes plastic to be mechanically weakened, allowing it to tear and puncture very easily. Also, some of the carbon materials make black marks on equipment, and particle contamination prevents its use in clean room applications. In addition, the possibility of corrosion by carbon due to galvanic action is always a problem to be considered.

Carbon-filled materials are electrically conductive, which must be considered for operator safety when working with electrically hot equipment. Electrically powered equipment must be insulated from conductive tabletop material or the tabletop will no longer have a soft ground, but be hard-grounded through the equipment.

ANTISTATIC MATERIALS

Nonconductive materials (plastics) can be made antistatic in two ways, by spraying an antistatic solution onto their surfaces, or by adding an

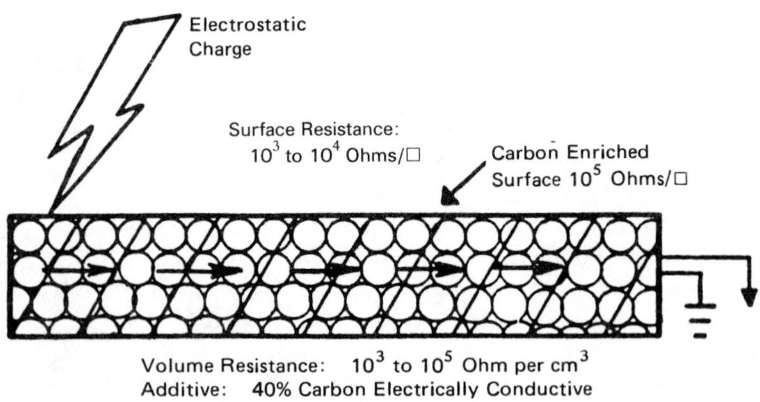

Figure 6-4. Carbon-loaded material.

antistatic agent internally to the raw material itself. There are a number of excellent antistatic agents on the market. When applied to the surface of plastic materials, they lower the surface resistance to the point where the material becomes conductive enough to prevent the generation of static electricity. Also, the application of an antistatic material to the surface of a plastic material will cause the former to drain off any static electricity, when connected to ground.

Antistatic materials are produced by dispersing an organic antistatic additive throughout a base plastic such as polyethylene. Being hygroscopic, the antistatic addititive, continuously migrating to the surface, causes the formation of a moisture layer on the host plastic by drawing water vapors from the air.

The moisture layer reduces friction and the degree of contact between the plastic and other surfaces. This in turn reduces triboelectric charge generation, which results from the contact and separation of materials. In addition, provides sufficient conductivity for relatively slow-charge redistribution and dissipation, when the material is grounded. This would happen, for example, if the material were in contact with a grounded, static-protective work surface, or if it were handled by personnel wearing a grounded wrist strap. The conductivity is not sufficient to produce effective shielding from external electrostatic fields, however. Antistatic material properties can also be obtained by periodically treating the surface of a host material with a topical antistatic.

Antistatic materials provide minimization of charge generation and retention when used in lieu of insulating, common plastics for tote boxes, kitting containers, bins, and other production aids. Not only are insulating plastic items efficient charge generators, they retain the charge even when grounded. However, antistatics do not provide effective shielding of external electrostatic fields.

During the manufacturing of antistatic materials, an antistatic solution is added to the raw material. After approximately 24 hours, the antistatic additive migrates to the surface and absorbs moisture from the air, lowering the surface conductivity to the point where it is impossible to generate static electricity by friction. It is the ability of this material to absorb moisture from the air that explains its performance. The moisture layer itself makes the plastic antistatic.

When the moisture layer is removed by using freon and clean cloth to wipe off an antistatic work top, the plastic is no longer antistatic. It will require 24 hours for the antistatic agent to remigrate to the surface, absorb another layer of moisture, and return the plastic to its original antistatic condition. If you clean antistatic work tops with a freon-saturated cloth at the end of each shift, you are defeating the very purpose for which you purchased the antistatic material. If a good antistatic spray or solution is used instead of freon, the antistatic tabletop will never lose its ability to absorb moisture and will always be antistatic.

DIP carriers, tote boxes, storage trays, shipping bags, and foams are available in antistatic plastic. Antistatic DIP carriers are often used and reused until all the antistatic agent is gone. This is false economics: the antistatic DIP carrier has been found to have more problems than any other antistatic device.

Some DIP carriers are made antistatic by use of an antistat spray or solution. Some of these external antistatic agents can be removed easily by continuous handling, causing the surface resistance to rise to the point where the DIP carrier will charge and store static electricity.

During spraying, as the antistatic solution leaves the aerosol can through the nozzle, the freon or propellent will evaporate immediately because of expansion. The alcohol, fire retardant, and antistatic agent all spread onto the surface. Soon the mild alcohol will evaporate, followed by the fire retardant, leaving only antistatic material on the surface to absorb moisture to cause the material to become antistatic. If the surface sprayed is porous, it will absorb the antistatic agent and will last a reasonably long period of time. However, if the surface is non-porous, the antistatic agent can be removed by continuous handling.

TOPICAL ANTISTATS

The topical antistat is a liquid chemical that may be sprayed or wiped on static-generating material surfaces so as to render them static-free. It is only a temporary fix, as the antistat is water-soluble and can also be abraded off. Topical antistats consist basically of a hydrophilic and are dependent on the relative humidity of the surrounding environment. They function by absorbing moisture to produce surface conductivity. Surface resistances vary from 10^7 to 10^{12} ohms per square, depending on the humidity.

There are many topical antistats available, but they should not be used unless they meet eight requirements. They should

1. Perform ionically.
2. Inhibit bacteria growth and fungus (EPA register and performance documented).
3. Be nontoxic by ingestion, inhalation, or dermal application.
4. Be nonflammable.
5. Be stable over long periods (shelf life before and after application).
6. Not a primary skin or eye irritant.
7. Have their performance documented by primary manufacturers.
8. Be cost effective.

In many cases antistats promote the growth of fungus and bacteria and may be a major eye and/or skin irritant. This last statement is true even if the antistat has an EPA rating for its concentrated form. The EPA rating must be for the actual application mixtures, to guarantee its safe use by personnel.

Conductive Paints or Coatings

Conductive paints or coatings typically are latex paints or acrylic coatings applied to the floor with a brush or roller. They range in resistance from 5×10^3 to more than 5×10^5 ohms per square. In general, their static performance is not well documented, although static decay times are less than 0.1 second per Federal Test Method 4046.

The major advantage of these materials is that they can provide environmental rather than localized static protection and can be readily applied in existing facilities. On the other hand, conductive paints and coatings have several limitations. They are usually carbon-loaded and messy to work with. They are subject to wear, flaking, and chipping and are not applicable for clean rooms. They require frequent reapplication and, over a period of time, can become expensive compared to some alternatives.

Topical Antistats

A more familiar product in the electronics industry is the topical antistat, a liquid consisting of two components: an antistatic mechanism and a carrier to transport it. Topical antistats function in three ways. First, they reduce the coefficient of friction to help minimize frictional static charge generation. Second, they increase surface conductivity to help dissipate a charge. Third, they interact with environmental factors to neutralize generated charges. Some absorb moisture from the environment to help create a conductive layer on the substrate. Others rely on an ionic interaction between the environment and the antistat to help neutralize the charge.

Because most antistats are diluted for use with water, they are easily applied to floors in the normal cleaning process. In addition, they provide environmental static protection. They generally are less expensive per application than conductive paints or coatings.

Topical antistats also have their limitations. First, because they're not principally intended for floors, scuff resistance and glossy appearance are sacrificed. Second, because most are soluble in both water and solvents, it's easy to accidently remove the antistat from the floor. Third, continual reapplication is necessary, usually more frequently than the manufacturer may suggest. Fourth, there seems to be variation in the features, quality, and performance among the various brands. For example, some are hygroscopic and their performance is humidity-dependent. Some antistats may contaminate clean room environments while others may be acceptable for these applications. Some may cause skin or eye irritation when applied.

Conductive Materials

Conductive materials are constructed in a variety of ways. Each product has its own conductivity value and other equivalent circuit resistances and capacitances. The result is differing rates of charge dissipation and electrostatic shielding effectiveness for each. Volume-conductive materials are produced by loading a base plastic with carbon.

Surface conductive materials are produced by either depositing a continuous metal layer on the base plastic, or coating fiberboard with carbon-loaded conductive ink, or laminating conductive foil to a paperboard base. Conductive containers, on the other hand, can provide effective electrostatic shielding. The dynamic response of a conductive container is characterized by a charge that is sufficiently mobile for rapid redistribution in response to electrostatic fields introduced into the vicinity.

Charge on the container surface quickly terminates electric field lines and prevents transient fields in the interior from reaching a level that jeopardizes sensitive parts. But conductive containers do not provide minimization of charge generation. They will retain the charge until grounded, at which time there is rapid charge dissipation.

SHIELDING

Since the early 1800s, when Michael Faraday conducted electrical experiments, it has been a well-known fact that field-induced charges caused by induction and polarization can be prevented from affecting an object by enclosing the object in a closed, highly conductive container. A container of this type has become known as a "Faraday cage."

The degree of shielding effectiveness of a Faraday cage is a function of the conductivity of the material from which the container is made. The more conductive it is, the more it will shield.

To illustrate the Faraday-cage effect, we may use a stryrofoam coffee cup and a sealed can containing coffee. The cup will be a positively charged object emitting a field. Because the can is sealed and made of metal, it is a very effective Faraday cage. The coffee is what we are trying to protect. As the cup approaches the coffee can, for a split second the electromagnetic field from the cup cuts through both the can and the

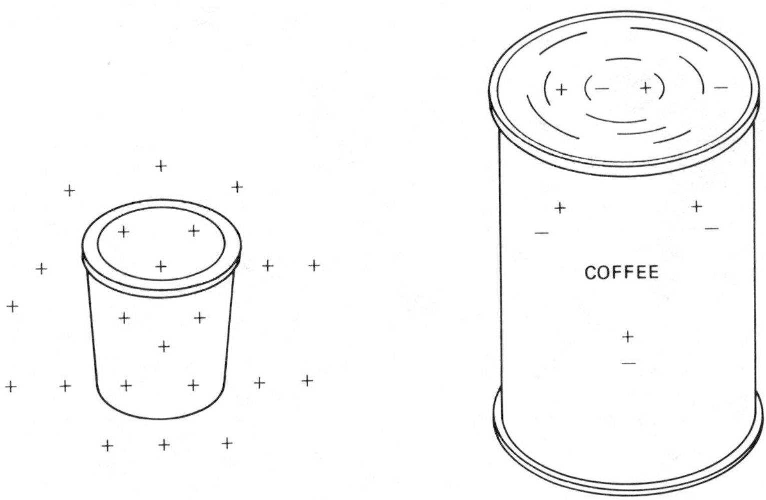

Figure 6-5. The Faraday-cage effect: the coffee can is isolated from the electromagnetic field of the styrofoam cup.

coffee inside, causing polarization of the can and the coffee to begin (see Figure 6-5.)

Because the can is made of highly conductive steel, and because good conductors have highly mobile electrons, redistribution of the mobile electrons in the can takes place almost instantly (see Figure 6-6). This causes the surface of the can to become polarized, but the coffee inside the can will remain neutral. The field force will travel around the can on the outside surface only. Once the can is polarized, the coffee inside the can will be unaffected by induction or polarization. In Figure 6-6 we see that the cup is charged (+5 −0 = +5), while the can, although polarized, is still electrically neutral (+5 −5 = 0).

Another law of physics known as Gauss's Law states that "the net outward or inward electrostatic field flux density through any closed surface is equal to the net positive or negative charge enclosed by that

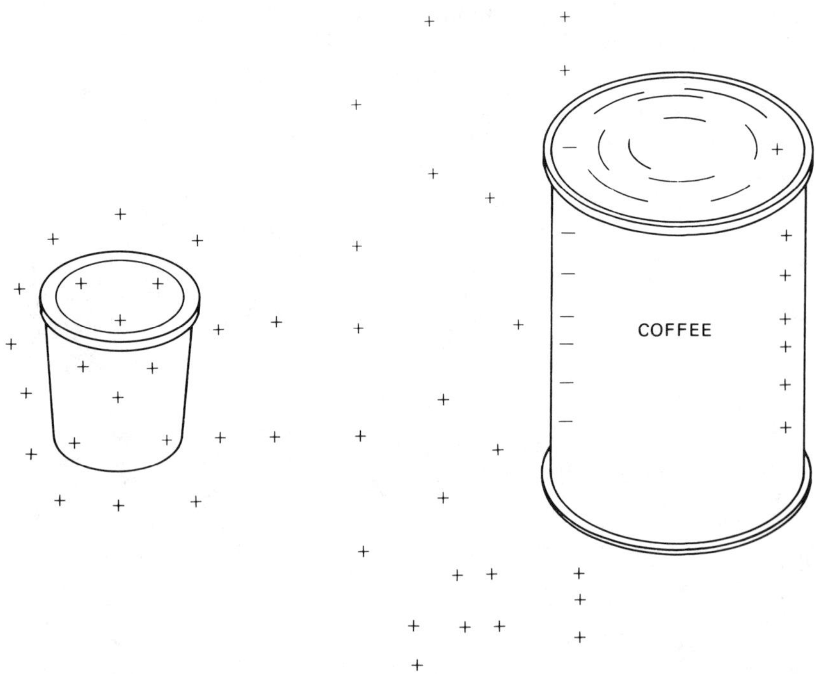

Figure 6-6. The Faraday-cage effect: the coffee can exposed to the electromagnetic field of the styrofoam cup becomes polarized.

surface." The coffee can is a closed container containing equal positive and negative charges, therefore, regardless of the charges' distribution, no net flux (electromagnetic field) will be found inside the can. The coffee is shielded from both inductive charging and polarization.

In shielding a sensitive electronic device, a shielding material must be able to realign its electrons extremely fast, lest the device be damaged. The more sensitive a device, the higher the level of shielding effectiveness it requires.

BASIC BAGS

One useful form for surrounding sensitive parts is a bag constructed from static-protective materials. Bags are made with either antistatic or conductive materials, or with layers of each in the same product (Figure 6-7).

Antistatic bags use polyethylene sheet material folded at the bottom and sealed on the sides, or tubular materials with a seal on the bottom only. Some types have a zipper lock at the top. However, a common method of bag closure is simply folding over the bag top and perhaps securing it with an adhesive-backed static warning label.

These bags are highly transparent, permitting personnel to examine their contents without removing them from the bag. Unnecessary opening of protective containers and removal of contents increases the probability that sensitive items will be subjected to damaging electrostatic

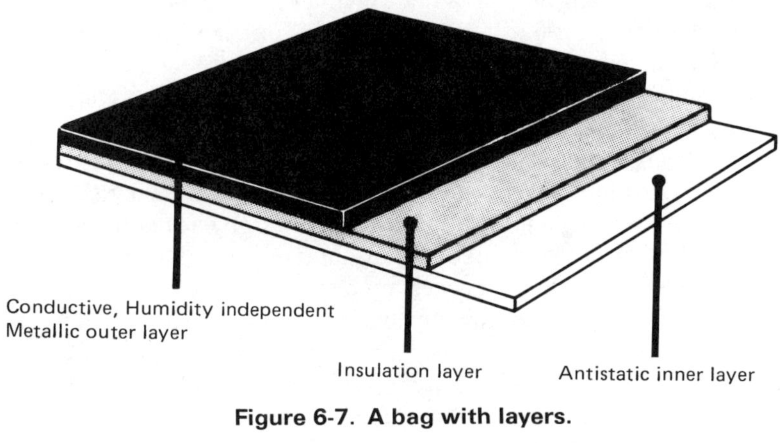

Figure 6-7. A bag with layers.

fields. The bags' properties depend on the ambient humidity. Antistatic properties will degrade over time, depending on extent of use.

Volume-conductive bags are usually constructed of carbon-loaded polyethylene. Since antistatic material is not used, charge generation is not minimized. The charge is rapidly dissipated as generated, if the bag is grounded through the work surface or the personnel handling the bag.

Surface-conductive bags provide some degree of transparency with their layered construction. The inner layer is heat-sealable, antistatic polyethylene; the second layer is polyester; and a continuous coating of nickel or aluminum forms the third layer. A thin protective film usually covers the metal layer.

Since a sensitive part within the bag will contact the polyethylene rather than the conductive layer, charge generation must be minimized by using antistatic material. Continuity of the conductive layer must be maintained in at least one direction around the bag to provide electrostatic shielding. Personnel must not crease and subsequently crack the metallic layer along a line that disrupts this continuity around the bag.

INTEGRATED PROTECTION

One protection element not yet considered is mechanical cushioning. This protection is provided by foam or bubble cushion, which is available in pouch and bag form as well as in sheets for wrapping sensitive items. A bubble pouch provides more trapped air space for increased cushioning and has a smooth inner surface for easy insertion of items (see Figure 6-8). Bags with layers of antistatic cushioning and conductive material that provide electrostatic shielding are available.

There is now greater recognition of traditional packaging considerations along with the static protection.

Integrating protective functions can provide several advantages over the approach that uses separate protective materials. First, labor is reduced by replacing a multistep with a single-step packaging process. Second, the variety of protective products in inventory is reduced. Third, one is assured that all necessary elements of protection are used.

Sensitive integrated circuit devices are protected in shipment by static-protective tube containers. Conductive DIP tubes are aluminum or carbon-loaded material (Figure 6-). Conductive tubes have the disadvantage of being opaque, so a small viewing slot is provided. Most

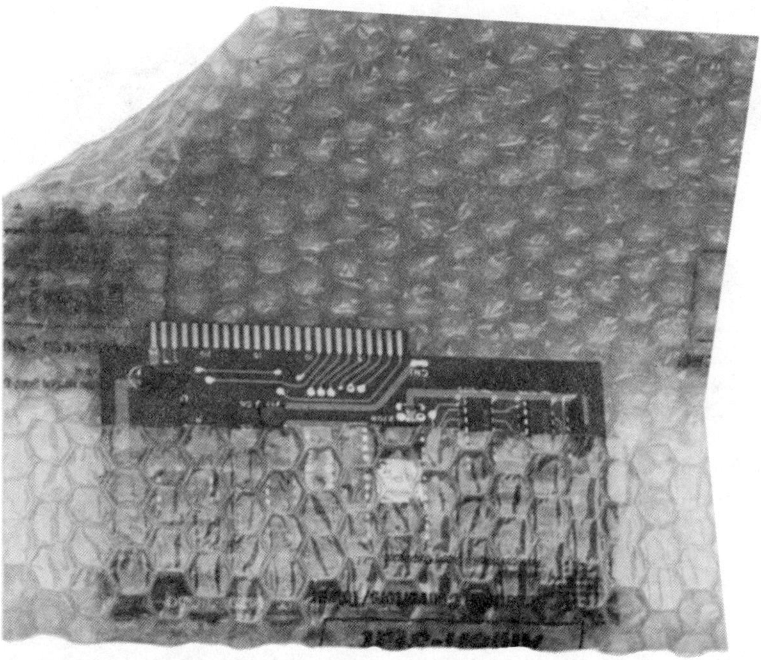

Figure 6-8. PCB protective bags, including carbon-loaded, antistatic, and bubble-cushioned.

users prefer antistatic tubes because of their transparency. Companies receiving devices in a DIP tube want to see the contents to detect bent leads before loading an automatic machine.

Tote boxes, kitting containers, and bins meet the needs of higher-level assembly operations (Figure 6-9). Use of conductive tote boxes is prevalent in storage applications. Antistatic containers predominate where there are frequent parts-transfer operations to and from the tote.

A complete electrostatic shield is formed with conductive tote boxes only when a lid is used. The type of lid is important. A lid with a lip that fits over the sides of the tote box, rather than one that sits within the top of the box, provides a better electrostatic shield and superior mechanical properties for handling and stacking.

Some firms extend application of existing insulating tote boxes to handling sensitive parts by using conductive liners. The insulating

Figure 6-9. These conductive containers can be partioned and expanded to fit many sizes of PCBs; they also can be adapted to robotics equipment. (Picture courtesy of Conductive Containers, Inc.)

plastic can generate and retain a charge, even when on a grounded, conductive surface. The insulating plastic also prevents charge dissipation from the conductive liner through a ground path. Treatment of the tote box with a topical antistatic on a periodic basis is also needed.

MATERIAL FLOW

Tote boxes for use with automated storage and retrieval systems, or with robots in the material-handling system, must provide flexibility in configuration and stable dimensions with close tolerances.

Protection of static-sensitive parts is one element in the overall concept of material flow in the factory. Good manufacturing systems which integrate the flow of materials throughout the factory result in less double handling and inappropriate storage of sensitive parts. This reduces the probability that sensitive parts will be damaged and also improves manufacturing efficiency.

Restricting components of the handling system to limited areas of operation multiplies the number of handling events. With compatibility in the handling equipment, linked movement of materials throughout a plant whenever parts must be transferred between containers by personnel who may carry a static charge is minimized.

The work surface, containers, and chair covering should use static protective materials in their construction. Dispensing and hanging rails for small-parts subcontainers and tote boxes should be grounded, just as the work surface and chair are.

Sensitive small parts should be prekitted in static-protective subcontainers in the receiving area. Placed in compatible-sized static-protective tote boxes, the subcontainers flow through the transport and storage steps with no transfer operations between tote boxes. Upon reaching a static-free work station, the subcontainers are slipped onto a grounded dispensing rail, where parts are accessible to personnel wearing wrist straps grounded through a connection on the work surface. The sensitive parts remain in the same static-protective container from the receiving area until they are dispensed at the work station. A similar process can be established for PC boards using static-protective board holders in a tote box, which can be hung from a grounded rail at the work station (Figure 7-5).

This system of getting all parts, sensitive or not, into containers early helps restrict vendor packaging material to a small area. Incoming insulating, common plastic, and particularly loose-fill and foam- or bubble-wrap cushioning that can generate and carry a charge, are confined to one area in the facility.

Mixing insulating plastic and conductive totes and bins can undermine the static-protective scheme. Using insulating plastic containers for nonsensitive parts brought to a static-controlled work area provides a source of possible damage to sensitive components. And even though conductive bins housed in nonconductive tote boxes are prevented from dissipating charge to ground, they can discharge through grounded personnel reaching into the bin, causing damage to parts in the bin.

Cardboard containers for shipping are constructed with conductive properties by application of a carbon-loaded coating or laminating aluminum foil to the cardboard. The conductive coating technique includes another outer coating to prevent sulphur in the fiberboard from contaminating device leads and affecting the soldering process.

EVALUATION OF ALTERNATIVES

How can one systemically evaluate the numerous alternatives offered by the combination of materials and container forms? By taking the following requirements into account, normally in this order: mechanical requirements, electrostatic protection requirements, and financial considerations.

The central concept of static control is to minimize charge generation in the first place. When it comes to considering static-protective materials, then, the use of antistatic materials in a controlled environment, and a combination of antistatic and conductive materials in an uncontrolled environment, are recommended.

For uncontrolled environments such as shipping, one needs conductive material between antistatic materials on each side. The outer antistatic material prevents charge from being generated and controls discharge from charged personnel. Ungrounded, charged conductive material, if used as the outer layer, can discharge to grounded personnel. Direct discharge to uncharged, conductive material from ungrounded, charged personnel can also cause damage to parts. Senstitive devices surrounded by antistatic material alone can be damaged by electostatic fields that may exist.

ECONOMIC CONSIDERATIONS

One of the first decisions to be made about the purchase of a static control work surfce is whether or not the user needs new benches for an expansion or if existing benches can be upgraded. If new benches are needed, the decision between a static control laminate or a table mat easily favors a laminate.

If the user currently has a facility with work benches in good condition, but without static control surfaces, it may be most cost-effective to use table mats. The mats can be placed on top of the existing work benches, avoiding the capital expense related to replacing the benches.

FUNCTIONAL CONSIDERATIONS: ELECTRICAL PROPERTIES

The primary purpose of the static control work surface, mat, or bench top is to drain static electricity. In all cases, the electricity is to be transferred from a charged object through the work surface to a connection that leads the charge to a zero potential ground. The most important properties of these products are their electrical characteristics. There are four electrical criteria to be considered: surface resistivity, surfce-to-ground resistance, decay time, and volume-versus-surface conductivity.

Surface Resistivity

Surface resistivity is a measure of the ability of a material's surface to conduct electrons. It is a property that is independent of the dimensions of the material. The units of this property are ohms per square. In order for a material to conduct the electricity away from a charged body along its surface, the surface resistivity must be below a certain value. However, in some systems the charge may be conducted through the top layer to a conductive sublayer and then to the ground connection, so that this property becomes less ciritical.

There are three major categories of static control materials, based on their surface resistivities. The resistivity ranges are:

Conductive	$<10^5$ ohms per square
Static-dissipative	10^5-10^9 ohms per square
Antistatic	10^9-10^{14} ohms per square

Surface-to-Ground Resistance

Surface-to-ground resistance defines the resistance of the actual electrical discharge path from the work surface to ground, and indicates how readily the charge can pass from its origin to ground. Its units are ohms.

Decay Time

Decay time relates the ability of the work surface to transfer the electrons from a charged body through the work surface to the ground connection. The decay rate varies inversely to the resistivity. Thus the greater the resistance, the slower the static charge decay rate. The choice of resistivity depends on the work environment, the types of electronic devices handled, and the rate of decay required.

Any analysis of electrostatic decay rate involving discharge through a laminate must include consideration of two equations. The first equation relates the total charge (Q) on a body to the voltage (V) on the body and the body's capacitance (C):

$$Q = CV \tag{1}$$

where

C = capacitance of the body to be discharged
V = the measured voltage on the body

The second equation relates the voltage on the charged body (V_t) after being dischargd to a ground, to the voltage before discharge (V_o), the resistivity along the discharge path (R), the capacitance of the body to be discharged (C_b), and the decay time (t), when V_t is measured.

$$V_t = V_o\, e^{-t/RC_b} \tag{2}$$

where

V_t = voltage after discharge at time t, volts.
V_o = voltage before discharge occurs, volts.
t = time at which V is measured, seconds. Mil-B-81705 requires a material to dissipate a 5,000-volt static charge to ground within 2.0 seconds.
R = electrical resistance along the discharge path, ohms.
C_b = the capacitance of the body being discharged, picofarads. C_b is typically 50-520 pf for a human and 50-150 pf for a conductive tote box.

Based on these two equations, it is possible to relate laminate resistivity properties directly to the rate of charge decay from a body. Rearranging the equation, R can be calculated:

$$R = t/(C_b \times \ln V_o/V_t) \tag{3}$$

This equation shows that a resistance of 4.35×10 is the maximum allowable to get a 0.1-second decay of voltage to 10% of its initial value with $V_o/V_t = 10$ and $C_b = 100 \times 10^{-12}$. This value should be kept in mind when comparing the various static control materials.

Volume-versus-Surface Conductivity

With some materials the electrical properties are homogeneous throughout and the material is volume-conductive. Volume-conductive materials have no advantage over surface-conductive ones, if the surface-conductive layer has sufficiently high wear. One disadvantage is the possibility of having a parallel path to ground that bypasses the current-limiting safety resistor.

Abrasion resistance. The life cycle expense of a mat or tabletop consists of three parts: initial cost, maintenance cost, and useful life. The relative initial costs of each of the products will have to be determined by getting quotes from various vendors. The maintainability of each product will be discussed relative to their basic types and will be part of their solvent-resistance and stain-resistance tests. The useful life of the products has been quantified in this program by use of the MEMA test for abrasion resistance. This test measures the relative ability of the product's surface layer to withstand sandpaper abrasion under well-defined circumstances.

Solvent resistance. Samples of the candidate materials should be tested with drops of solvent residing on their surface and soaked with solvents, varying from the extremes of deionized water to trichloroethane, to determine their performance.

Stain resistance. Staining agents should be poured on the surface of each material, then standard commercial cleaners should be used to clean it off. The presence of any residual stains should be noted.

Heat resistance. This property is based on the material's ability to withstand hot solder dropping on it, or a soldering iron placed on its surface.

AESTHETIC CONSIDERATIONS

Since "beauty is in the eye of the beholder," the basic descriptions of each type of product are given, as well as a general comment about their aesthetics. Of primary importance are glare and eye strain. Black and other dark color materials are often considered undesirable for work surfaces, because of their low contrast with the electronic parts to be handled on them.

MAGAZINES (SLIDES, TUBES, RAILS): ANTISTATIC TYPES

Magazines should be specified by one of the following classes and be tested for triboelectric charging by using the Faraday-cup test. In the testing of the magazine specimens, the limits should not exceed those specified for that class by this test.

Class I The charge shall not exceed 0.03 nanocoulombs per lead for any leaded device containing a MOS type structure.

Class II The charge shall not exceed 0.05 nanocoulombs per lead for any device containing a bipolar structure.

Class III The charge shall not exceed 0.01 nanocoulombs per lead for any device considered nonleaded (e.g., a chip carrier or bumped package) which contains a MOS-type structure.

Class IV The charge shall not exceed 0.02 nanocoulombs per lead for any device considered nonleaded (e.g., a chip carrier or bumped package) which contains a bipolar structure.

These magazines should meet one of the following material requirements:

1. Have a resistance end to end of less than 1.0×10^6 ohms, with the shielding surface not necessarily exposed.
2. Have a shielding layer with a surface resistivity of less than 1.0×10^5 ohms per square as measured per ASTM-D-257 or the equivalent.
3. Have a volume resistivity of less than 1.0×10^4 ohm-centimeters as measured per ASTM D-991 or the equivalent.

Table 6-2. Maximum Values for Shielding Classes.

CLASS	V_o	V_p (max.)
I	10,000	100
II	5,000	100
III	10,000	500
IV	5,000	500
V	10,000	2,000
VI	5,000	2,000

The maximum measured test value for each shielding class should be that shown in Table 6-2, where V_o is the initial voltage on the 200 pf capacitor and V_p is the maximum acceptable peak amplitude voltage across the capacitive sensor as observed on the oscilloscope.

BAGS, POUCHES, AND SHEETS (NONCUSHIONING)

These materials should be evaluated by the following procedures.

Electrostatic Decay Time

Homogeneous packaging materials should have an electrostatic decay time of not more than 2.0 seconds as measured by the test apparatus of Federal Test Method Standard 101, Test Method 4046, or the equivalent, when measuring both an applied positive and negative 5,000-volt charge dissipating to 50 volts after the material is grounded. The instrumentation shall be calibrated to have a deomonstrated precision and accuracy within a ± 10%.

Surface Resistivity

Surface resistivity of antistatic plastics shall be less than 1.0×10^{14} ohms per square, but greater than 1.0×10^{10} ohms per square on both surfaces, as measured in accordance with ASTM D-257 or the equivalent.

Conductive Types

Materials which are surface-conductive only should have a surface resistivity of less than 1.0×10^{5} ohms per square. Volume-conductive materials should have a volume resistivity of less than 1.0×10^4 ohm-centimeters.

Electrostatic-Shielding Types

The shielding surface should be electrically continuous. The surface resistivity should be less than 1.0×10^9 ohms per square or, in the case of volume-conductive materials, less than 10 ohm-centimeters per mil of thickness.

IDENTIFICATION

All static-controlled materials should be individually identified as electrostatic types by marking, embossing, printing, labeling, or other suitable means, to prevent use of other similar but nonelectrostatic shielding materials where electrostatic shielding properties are desired. They should be marked with the manufacturer's name, lot number, and date of manufacture by month and year (e.g., 8/82).

7. MATERIAL CONTROL AND TRANSPORTATION

ESD causes many types of damage. The charge generated by the insertion and removal of a loaded board from a bag can be strong enough to do great harm. And static electricity also does some dirty work—literally—by creating a "magnetic" attraction between charged airborne contaminants and surfaces of the opposite charge. These particles which cannot be brushed off, can cause circuit shorting and degradation.

In making and specifying the containers to protect boards from these effects, four widely accepted, though certainly not rigid, categories are used. *Conductive* materials are generally those that exhibit surface resistivities of 10^5 ohms per square or less and include metals, bulk conductive plastics, wire-impregnated materials, and conductive laminates. (Bulk conductive material, with volume resistivity of 10^4 ohms/cm, needs a minimum thickness of 0.1 cm; if it is any thinner, it falls into the static-dissipative category.) *Static-dissipative* materials, with surface resistivities between 10^5 and 10^9 ohms per square, include the same materials just listed, only thinner, or with finer or sparser wire or wire mesh, or with higher volume resistivities. *Antistatic* materials, with surface resistivities between 10^9 and 10^{14} ohms per square, comprise hygroscopic antistatic materials, some melamine laminates, high-resistance bulk conductive plastics, virgin cotton, cellulose-based hardboards, wood, and paper products. Very thin static-dissipative or conductive materials also fall into this category. *Insulating* materials like polyethylene have surface resistivities greater than 10^{14} ohms per square; these are prime static generators and hold charges for long periods of time. The amount of time it takes for the voltage to decay to a certain level on a charged material—called decay time—is an indirect measure of surface resistivity.

Products made to protect moving printed circuit boards, during assembly and when finished, come in many shapes and materials. Polyolefin and nylon are often impregnated with carbon to increase their conductivity. Antistatic polyethylene or nylon come with antistatic additives incorporated directly into the resin. A microscopic layer of the additive is generated onto the plastic film surface; when removed or worn off, this layer migrates to the surface from a reservoir in the bulk film. The additive, a hygroscopic surfactant, attracts water from the atmosphere, providing a slightly conductive surface.

Combinations of materials are also appearing in the static protection market: a layer of antistatic nylon, a conductive carbon grid, and an antistatic copolymer. Since the carbon is buried under the nylon, it does not shed particles. Another hybrid is the transparent bag which features an antistatic polyethylene interior, a metallic outer layer for shielding, and a polyester layer.

Topical antistats include detergent substances which wet the host surface. Since they are hygroscopic, they are most effective at high humidities. Other antistats, which are not humidity-dependent, reduce static generation by increasing surface lubricity and conductivity. Topical antistats work mainly to reduce the ESD caused by friction and to provide some conductivity for charge bleed-off. Since these antistats wear out, surfaces covered with them must be checked periodically with a static detector.

Transparency is preferable, since it permits board inspection without direct handling. Heat sealability means a bag can be firmly closed and will protect against dirt and humidity. Resistance to solvents is important, if cleaning is necessary. The less sloughing of particles, the smaller the possibility of damage to board components from abrasion and creasing. The material should exhibit flex resistance, tensile strength, foldability, elongation, and impact strength. Antistatic products are generally translucent or transparent pink, are inexpensive, need periodic tests to determine functioning, and require upkeep. Also, they depend on relative humidity.

Carbon-loaded products, which slough carbon particles, are not fit for clean rooms. The humidity that antistats require can cause corrosion and other component damage. Transparent conductive shielding bags, if creased too much, can lose metal continuity, destroying their Faraday-cage effect. And topical antistats sometimes promote fungal and bacterial growth and can be an eye and skin irritant.

A PCB with a slight overall charge can be damaged by sudden localized grounding when touched to a conductor. Far more damage is done to sensitive devices by exposure to plastic and nonconductors which can induce charges on board wiring and can discharge repeatedly.

Since antistats produce neither electrostatic charges nor sparks, they are ideal for handling and transporting ESD-sensitive boards.

BASIC BAGS

One useful form for surrounding sensitive parts is a bag made with antistatic or conductive materials as well as layers of each in the same product (see Figure 7-1). These bags are highly transparent, permitting personnel to examine contents without removing them from the bag. Unnecessary opening of protective containers and removal of contents increase the probability that sensitive items will be subjected to damaging electrostatic fields.

The properties of antistatic bags depend on the ambient humidity; additionally, antistatic properties degrade over time, depending on the extent of use. Conductive bags are usually constructed of carbon-loaded polyethylene. Since antistatic material is not used, charge generation is not minimized. Charge is, however, rapidly dissipated as generated, if the bag is grounded through the work surface or personnel handling the bag.

Surface-conductive bags provide some degree of transparency with their layered construction. The inner layer is heat-sealable, antistatic polyethylene, the second layer is polyester, and a continuous coating of a conductive material forms the third layer. A thin protective film covers the metal layer. Since a sensitive part within the bag will contact the polyethylene rather than the conductive layer, charge generation must be minimized by using antistatic material. The polyester provides a strong mechanical and dielectric base for the metallic layer.

Continuity of the conductive layer must be maintained in at least one direction around the bag to provide electrostatic shielding. Thus personnel must not crease and subsequently crack the metallic layer along a line that disrupts this continuity around the bag. Manufacturers of these bags claim that the inner antistatic layer is not as affected by humidity changes and wear, as are the properties of completely antistatic bags. The shielding layer is not affected by humidity.

INTEGRATED PROTECTION

Another protection element is that of mechanical cushioning. This protection is provided by foam or bubble cushioning, which is available in pouch and bag form as well as in sheets. A bubble pouch provides more trapped air space for increased cushioning and has a smooth inner surface for easy insertion of items. An antistatic cushion bag of multiwall construction with a zipper closure is available. This bag has layers of smooth antistatic polyethylene on each side of the antistatic cushioning layer. The zip closure is formed in extensions of the smooth polyethylene above the cushion.

Bags with layers of antistatic cushioning and conductive material that provide electrostatic shielding are available. The first or outermost layer is smooth antistatic polyethylene, the second is conductive polyethylene, the third is antistatic cushioning, and the fourth or innermost is again smooth antistatic polyethylene.

There is now greater recognition of traditional packaging considerations along with the static protection.

Integrating protective functions can provide several advantages over the approach that uses separate protective materials. First, labor is reduced by replacing a multistep with a single-step packaging process. Second, the variety of protective products in inventory is reduced. Third, one is assured that all necessary elements of protection are used.

Sensitive integrated circuit devices are protected in shipment by static-protective tube containers. Conductive DIP tubes are aluminum or carbon-loaded material. Conductive tubes have the disadvantage of being opaque, so a small viewing slot is provided. Users, according to Charleswater Products, prefer antistatic tubes because of their transparency. Companies receiving devices in a DIP tube want to see the contents, so as to detect bent leads before loading an automatic machine.

Tote boxes, kitting containers, and bins meet the needs of higher-level assembly operations. Use of conductive tote boxes is prevalent in storage applications. Antistatic containers predominate where there are frequent parts transfer operations to and from the tote.

A complete electrostatic shield is formed with conductive tote boxes only when a lid is used. The type of lid is important. A lid with a lip that fits over the sides of the tote box, rather than one that sits within the top of the box, provides a better electrostatic shield and superior mechanical properties for handling and stacking.

Some firms extend application of existing insulating tote boxes to handling sensitive parts by using conductive liners. The insulating plastic can, however, still generate and retain a charge, even when on a grounded, conductive surface. The insulating plastic, in addition,

Figure 7-1. A conductive bag to provide shielding from external electrostatic sources.

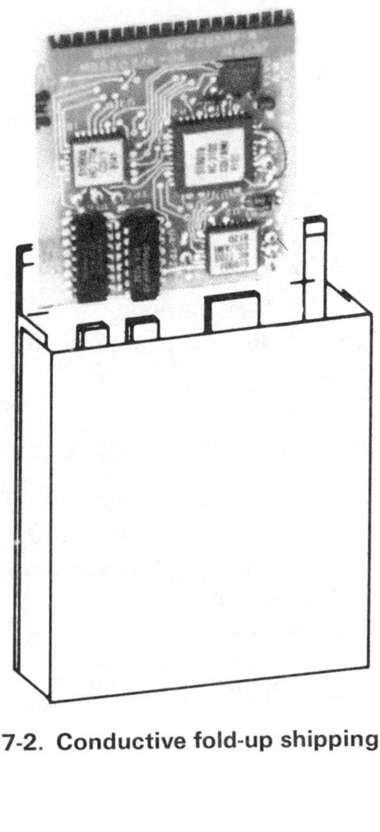

Figure 7-2. Conductive fold-up shipping carton.

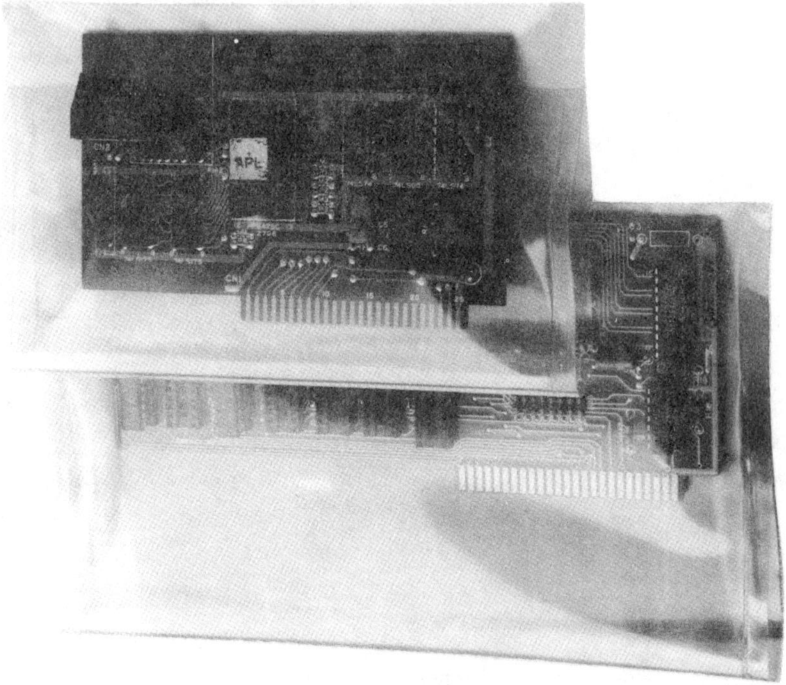

Figure 7-3. Antistatic see-through bags (pink poly).

prevents charge dissipation from the conductive liner through a ground path. Treatment of the tote box with a topical antistatic on a periodic basis is also needed.

ELECTROSTATIC DISCHARGE (ESD)—AN ECONOMIC PROBLEM

Electrostatic electricity is first and foremost an economic problem. If contamination and ESD are not causing losses in your plant, then there is no problem (even though static electricity is present). But if losses are occurring, a solution must be found that will produce a sound cost-effective operation. Often the problems encountered with ESD are solved on a piecemeal basis. The first vendor called appears with

Figure 7-4. Printed circuit rack.

his "cure-all," product which is immediately implemented and shows no tangible results. This becomes very costly.

Before approaching any solution, the user must address and clearly define the problem and generate a control plan. If required, an outside consultant should be carefully selected and brought in to provide this plan. Although many manufacturers of static control materials and

Figure 7-5. Tote box rack.

devices provide this service, an independent consultant that has no product to sell is recommended, since you will be his only interest.

Defining the need for a static control program should be based on the answers to the following questions:

1. The sensitivity of the components in use.
2. The physical areas to be protected.
3. The mobility of personnel and equipment.
4. The potential for management control.
5. The cost/benefit analysis.

Device degradation, which is responsible for the latent failure of equipment, decreases product reliability, increases field engineering cost, and can produce an additional circumstance that most companies

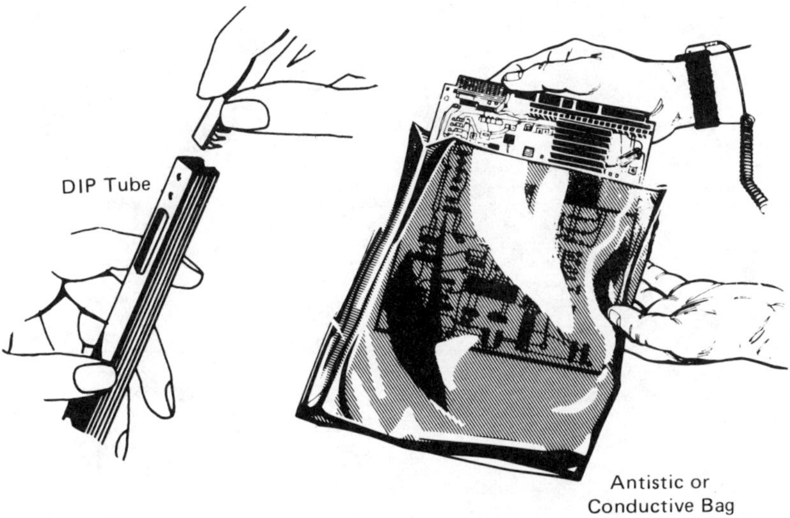

Figure 7-6. Component and subassembly protection.

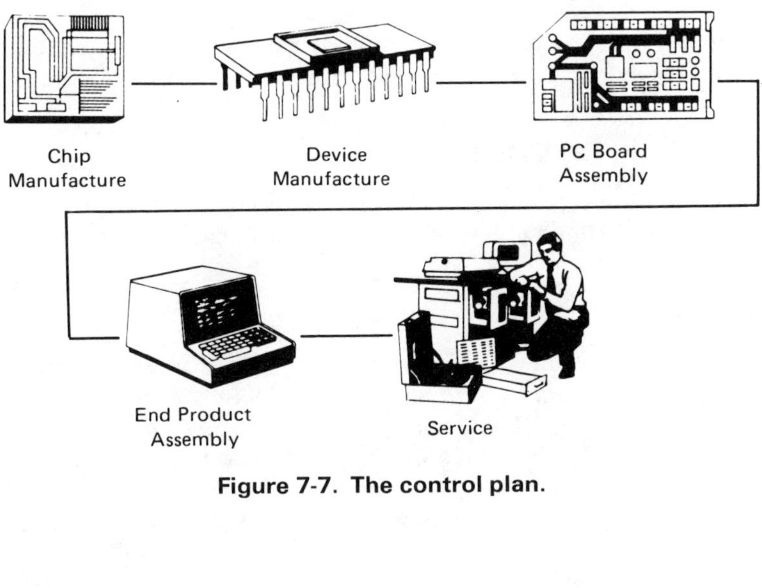

Figure 7-7. The control plan.

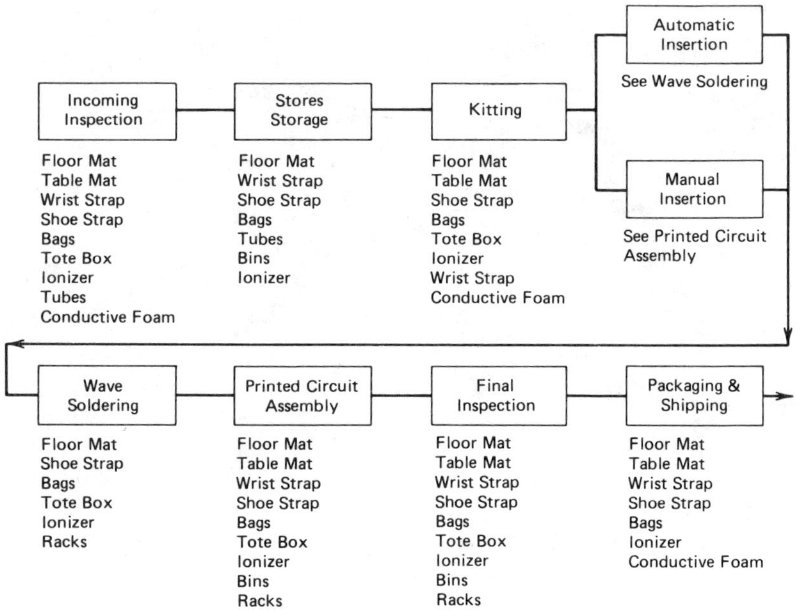

Figure 7-8. Typical static problem areas and their solutions.

overlook—a risk of liabilities arising from equipment malfunction or failure that jeopardizes human life or expensive and critical processes.

THE STATIC CONTROL PLAN

The basic control plan implements the materials and devices required in three well-defined areas: protection, grounding, and ionization.

Protection

All static-sensitive devices entering the operation must be protected from the receiving area through the manufacturing cycle, including testing, to shipping of the final product. Wherever possible, static-generating materials such as coffee cups, candy wrappers, plastics, and so on should be eliminated. Materials and equipment should be selected on the basis of their ability to prevent the possibility of holding a static charge on all surfaces or to drain static charges to ground. Personnel must be properly trained in how to handle static-sensitive parts and static-deterrent tools and equipment.

Grounding

The application of grounding systems must be carefully designed to ensure that the work station is continually cleared of all static charges. Current-limiting resistors must be installed to protect the operators from any potential electric shock. Methods must be implemented to allow static-sensitive components to drain sufficiently when placed on the grounded work mats before removal from transport containers, so as to eliminate the possibility of sudden discharge when removed from their containers. Grounding systems should be checked periodically to verify that there are no opens in the circuit.

The Elements to a Static Control Program

1. Protection—Any static control program must begin with the protection of devices entering your operation. Incoming static sensitive devices must be protected by "conductive" packaging materials.
2. Grounding—Grounding is essential to prevent a static charge from developing. Grounding cannot and will not trade off any static electricity already in existence in a nonconductive device. Any plastic device manufactured from a nonconductor cannot be neutralized by placing on a grounded, conductive surface. As mentioned previously, a nonconductor will not conduct static electricity; the charge remains or builds.

Ionization. When a static charge has developed or is in the process of developing on anything that is nonconductive, the only solution is ionization. Ionization can be used effectively to neutralize charges which already exist or to maintain a neutralized environment where charges are prone to develop.The use of grounding without ionization is tantamount to no protection at all. For example, a grounded operator coming in contact with a charged device (a device that has been stored in a nonconductive container) will, in too many instances, blow the device.

Summary

1. Your ESD control plan must begin at your vendor's plant. All sensitive devices should be shipped to you in conductive packaging material.
2. Receiving and inspection operations should be performed only in a grounded, ionized environment. No personnel should be allowed to unpack or handle sensitive devices in any other than this protective environment.
3. As your static-sensitive devices are transported throughout your operation, they should be packaged in a conductive container. Likewise, if you are utilizing hand carts, the wheels should be made of conductive material. If not, a drag chain can be installed to keep each metal cart grounded as it is moved throughout your plant. Automatic transfer devices, such as conveyors, should be augmented by ionizing equipment located at all transfer points.
4. Assembly stations should be totally grounded. The work station itself should be a conductive material. The operator should be required to wear a wrist strap and stand on a conductive floor mat. An ionizer should be used to neutralize all devices and materials coming into the work area. Special assembly equipment, such as rail assembly stations, must be covered with ionizers, and both the station and the operator must be grounded.
5. Process machinery must be grounded. Sensitive devices and PC Board assemblies entering and leaving processed equipment should be exposed to an ionization unit before and after processing. If a conveyor is used to transfer the device through the processed equipment, it should be protected with an ionizer to ensure that the conveyor stays static-free.
6. Finally, as the completed product is ready for shipment to either the end user and/or sister plant, it should be neutralized and then packaged in a conductive container. This packaging operation should be carried out in a complete work station including all grounding devices and ionizing equipment.

8. TESTING AND MONITORING

Service and maintenance personnel should use the same ESD precautions as prescribed for the manufacturing people in the previous chapters. On reaching the troubled unit, the repair person should place a floor or table mat on a convenient surface adjacent to the troubled unit. A ground cord from the mat to any reliable ground using an alligator clip should first be installed. Water pipes, unpainted equipment frames, electrical conduit, or the building frame usually provide an acceptable ground point. (See Figure 8-1.)

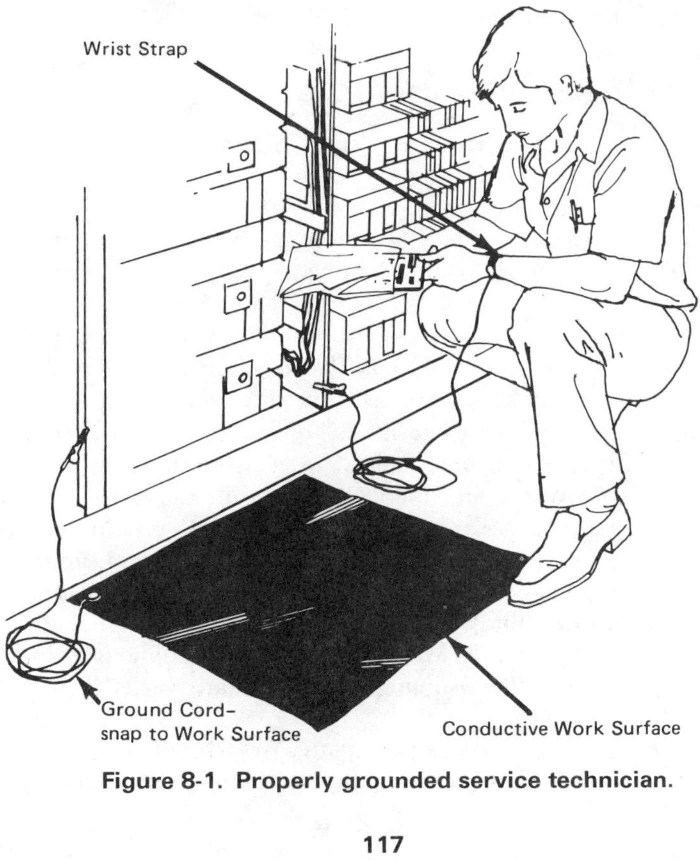

Figure 8-1. Properly grounded service technician.

A wrist strap properly connected to the mat (an alligator clip may be used on the ground strap, to connect it to the mat) must be worn. Contact between the wrist and the wrist strap need not be tight, but an extremely loose fit may not provide a good electrical connection and so should be avoided.

It is now safe to handle static-sensitive devices. If a printed circuit card is removed, it should be placed into a static-shielding bag while the repair person is still properly grounded. If a static-shielding bag is not available, the printed circuit board may be rolled up in the mat for transport to the service area.

If the suspect printed circuit board is to be repaired on site, it may be safely handled on the grounded mat. The grounded mat will protect it from static-laden carpeting or vinyl flooring. If the problems persist, the trouble may be user-caused, and placing a permanent mat in front of the equipment may be the answer.

SURFACE RESISTIVITY VERSUS RESISTANCE

The unit of measure used to determine the conductivity of barrier materials causes some confusion. Resistance is measured in ohms, while surface resistivity is measured in ohms per square. The accepted method of measuring surface resistivity is covered in test method ASTM-D257, published by the American Society for Testing and Materials.

Resistivity and conductivity are opposite terms: by simply measuring the resistivity a material has to the flow of electrons, the conductivity of the material can be determined. The simplest way to explain the difference between the two terms is to equate electrical units of measure with inches or square inches. To measure the resistance of a piece of copper wire, use an ohmmeter to measure the resistance from one end of the wire to the other. (See Figure 8-2.) To measure surface resistivity of a material, measure the resistance in all directions on the entire surface. Simply put, resistance measurements in ohms are similar to measuring the length of the piece of wire from A to B in linear inches, feet, or yards. Measuring surface resistivity is similar to measuring square inches, square feet, or square yards of a given surface. (See Figure 8-3.)

The unit of measure is ohms per square. We ask, "Ohms per square *what*?" We can answer this question with some simple electrical diagrams.

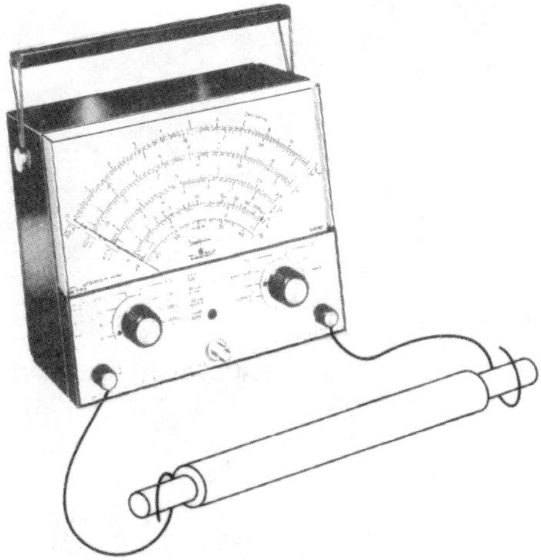

Figure 8-2. Measuring the resistance of copper wire with an ohmmeter.

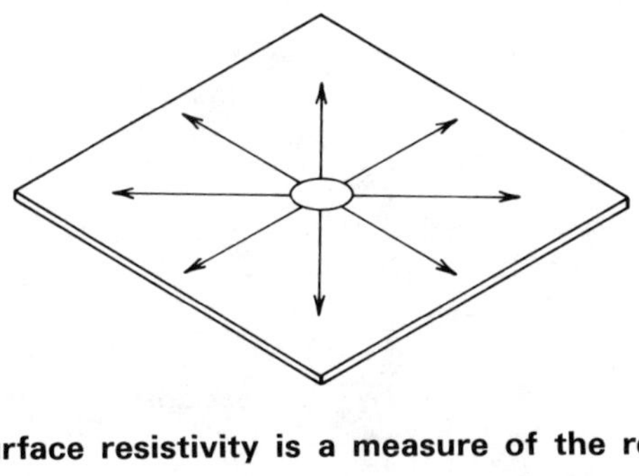

Figure 8-3. Surface resistivity is a measure of the resistance in all directions.

It makes no difference whether the sequence is in inches, feet, yards, or miles. Take, for example, a square piece of material that measures one foot on each side and has a resistance of 5 ohms from the left side to the right side of the square (Figure 8-4). If one adds a second piece of the same material to one side of this square, it is no longer a square (Figure 8-5), because the two pieces have been electrically connected in series. The resistance from the left side to the right side is the sum of the resistance of both pieces, so total resistance becomes 5 + 5 ohms or a total of 10 ohms.

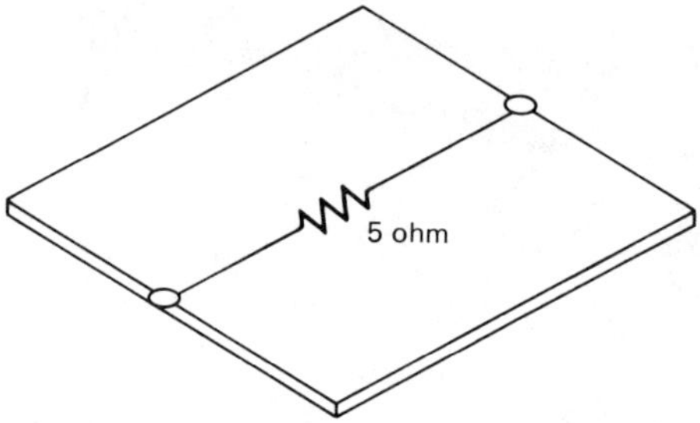

Figure 8-4. The resistance of the square is constant, regardless of dimension, feet, inch, yard, etc.

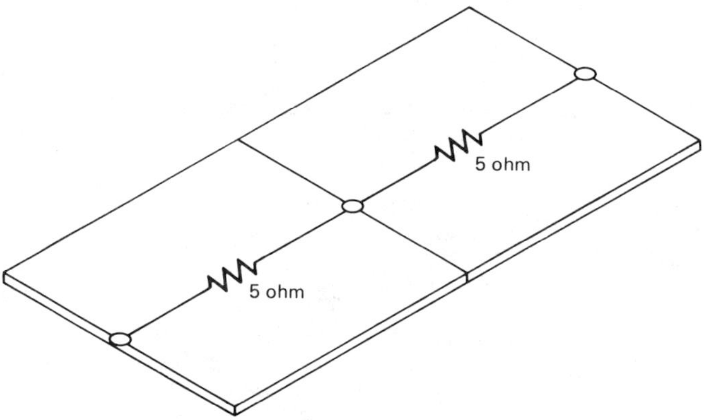

Figure 8-5. Two squares side by side become a series connection: 5 + 5 = 10.

Adding a $1' \times 2'$ piece of the same material to the edge of the other two pieces makes the measurement two feet on each side (Figure 8-6). This last piece creates a second parallel path for electrons to flow across the surface of the material. When resistances are connected in parallel, rather than in series, current flow is divided between two paths. Since the sum of the resistance in the upper path is 10 ohms, and the total resistance in the lower path is 10 ohms, each path carries half the electrons. Next, divide the resistance of each path by the total number of paths: $10/2 = 5$ ohms. This is the same result as when the square was only $1' \times 1'$.

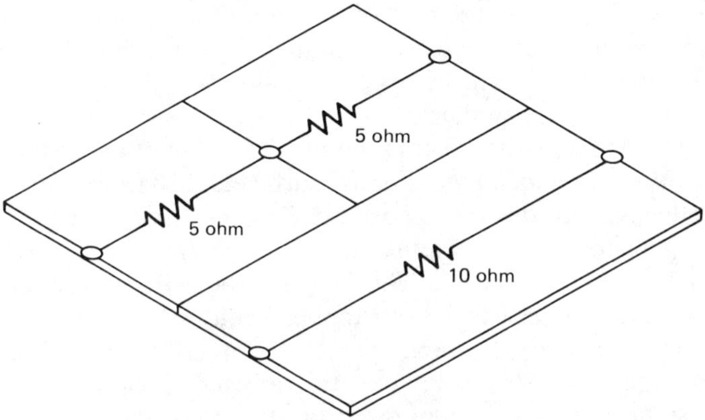

Figure 8-6. Addition of a third piece provides a second (parallel) path. Each path carries only half the current.

From these simple figures, it is evident that as long as the surface being measured is square, the actual dimensions become immaterial. Therefore it doesn't matter whether the square is a square inch, a square foot, or a square mile; the surface resistivity will be the same. The answer to the question "Ohms per square *what*?" is ohms per square *anything.*

MEASURING SURFACE RESISTIVITY/ CONDUCTIVITY

As already noted, resistivity and conductivity are opposite terms. If we need to know how conductive a material is, we measure its opposite, resistance. The more conductive a material is, the lower its resistance. The conductivity of a material is important when it comes to making a package that will protect a static-sensitive item. The more conductive a material is, the faster it will absorb and convey the charge to ground.

The unit of measure for electrical resistance is the ohm, just as the unit of measure for length is the inch. The ohm is a very tiny unit of measure, so it takes lots of ohms to stop the flow of an electrical current. To get an accurate measurement in ohms of the resistance of a given material, it takes some good equipment and a standard way of making the measurement. One method of measuring was developed by the American Society for Testing and Materials (ASTM) under their number D-257. A second method of measuring is described in the military specification for black conductive plastic film, from which black conductive bags are made (specification MIL-B-82646). Basically, the two methods are similar.

The test begins with a 25-cm (about 10″) sample of the test material placed on a flat, rigid, nonconductive surface. A test fixture having two knife-edged electrodes spaced exactly 2.54 cm (1″) apart is placed on top of the material to be measured (see Figure 8-7). A 5-pound weight is placed on top of the fixture. The two electrodes from the fixture are connected to a meter which both applies a voltage of less than 10 volts (DC), and measures the ohms resistance of the 1 inch of material between the electrodes. Since electrical resistance increases with the distance the electrical current must travel through, the two electrodes must be exactly 1 inch apart, and since the pressure of the contacts on the sample can change the measurement, exactly 5 pounds of weight is applied.

Notes:

1. The electrodes shall be metallic; the edges touching the specimen ground shall be flat, polished, and clean.

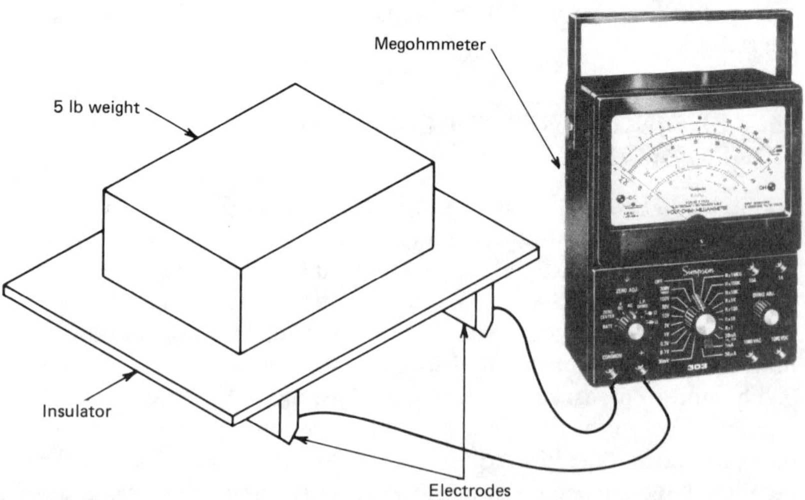

Figure 8-7. Military electrical resistivity measurement setup (MIL-B-82646).

2. The electrodes shall be permanently attached to an insulator, having minimum thickness of 0.64 cm.
3. The electrode edges contacting the specimen shall lie in the same plane.
4. The distance between the knife edges of the electrodes shall be $2.540 + .025 - 0.0$ cm.
5. The meter shall be an ohmmeter of suitable accuracy which applies a potential not greater than 10 volts at the electrodes (see 6.4).

According to the military specification, five readings are taken at random points on the sample, and the average of the five readings shall not exceed 300 ohm-cm. The formula used to make the calculation is as follows:

$$\begin{array}{c} \text{Measured resistance} \\ \text{(in ohms)} \end{array} \times \begin{array}{c} \text{Thickness of sample} \\ \text{(in cm)} \end{array} = \begin{array}{c} \text{Electrical resistivity} \\ \text{(in ohm-cm)} \end{array}$$

The test described is used to measure the resistivity of thin conductors.

VOLTAGE SUPPRESSION

When conducting a static audit in a plant, an understanding of voltage suppression is necessary. Most static meters operate by sensing the electromagnetic field associated with a static charge on an object. When two nonconductive objects with equal but opposite charges come together, one charge will balance the other, and the electromagnetic fields of both objects will be neutralized: no field will be present. Once the two nonconductive objects are separated from each other, however, the charge on each will reappear.

The instant the objects are separated, the meter will show that one still has a positive charge, and the other a negative. In the real world, a perfect balance of charges seldom exists, so field meters will usually give a reading. Where opposite charges exist, the meter will read the stronger of the two, reduced by an amount equal to the lesser. For example, if a styrofoam coffee cup with a 15,000-volt negative charge on its surface rests on a glass-top table having a 10,000-volt positive charge on the glass, the cup will indicate a negative charge of about 5,000 volts. But if you lift the cup into free space by means of insulated

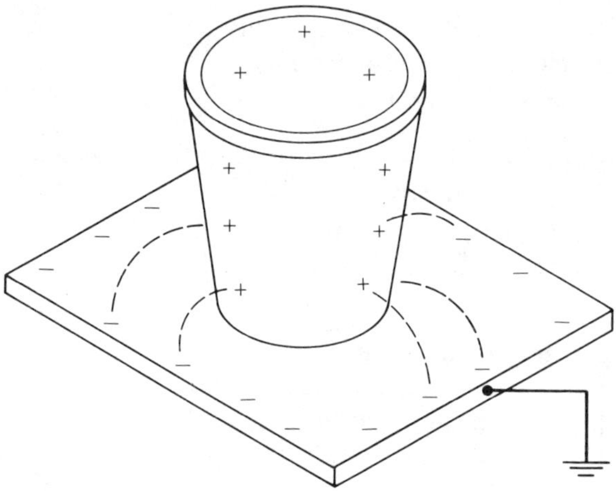

Figure 8-8. A positive charge on a styrofoam cup induces a charge of opposite polarity on a conductive mat.

tongs, the meter will read the full 15,000-volt charge on the cup.

Voltage suppression can also occur when a nonconductor is in contact with either a grounded or an ungrounded conductor. Placing a charged nonconductor on a grounded conductive surface will not remove the charge from the nonconductor. Figure 8-8 helps explain what takes place. Any time an object is charged, it emits an electromagnetic field. In this case, the positive field from the cup will induce a charge of opposite polarity on the grounded, conductive surface. The charge on the cup and the induced charge on the surface will at least partially offset each other.

If a field meter is brought near the cup while the cup is in contact with the surface, the field meter will register only the unneutralized charge on the cup. Once the cup is lifted from the surface with insulated tongs, the full charge will reappear on the cup, and the induced charge on the grounded surface will drain to ground.

Because charged insulators retain their charge, and voltage suppression can and does occur, don't be fooled by field meter readings that show small charges. To be safe, prohibit the use of all nonconductors at a static-safe work station.

MEASURING STATIC DECAY TIME

Another important property of a static discharge protective material is the speed at which it will dissipate a charge. Considering this in terms of a sponge makes it easier to understand. A good sponge will soak up a lot of water; a good static protection material will soak up a lot of electrical charge. Of the sponge, one asks, "How long does it take to dry out?" Of the static protective material, one asks, "How long does it take for the static to bleed off to the atmosphere or to ground?"

In the case of the sponge the thickness, the amount of water, and the humidity of the air around the sponge all affect how fast the sponge will dry. With a static protective material the thickness of the material, the amount of the static charge, and the atmosphere around the material all have a direct bearing on how fast the charge will dissipate.

To measure the speed at which a material will dissipate a charge requires sophisticated equipment. An Electro-Tech Systems Model 406B static decay meter, complete with environmental chamber, is recognized as an alternative for equipment specified in Federal Standard 101, Test Method 4046. Both the Electronics Industry Association (EIA) in its Interim Standard P.N. 1525, and the Reliability Analysis Center (RAC) in its recent report SOAR-1, mention the ETS, model 406B, as an accepted static decay test setup.

To test a material, at least nine samples of the material must be selected at random from a given batch of material. Three of the nine samples are placed in an oven maintained at 160° F ± 5° F for 12 days. Three more must be laid horizontally under a water bath for 24 hours before the test, and the last three must be maintained at 73° F ± 5° F in 50% relative humidity. The nine samples must then be placed inside the environmental chamber of the static decay tester for another 24 hours before the tests begin. The chamber of the tester must be maintained at 73° F ± 3.5° F with a dry condition of less than 15% relative humidity.

Each sample is charged by applying a 5,000-volt positive charge. The sample is grounded, and the meter outside the test chamber measures the time it takes to dissipate 100% of the applied charge. This process is repeated three times for each sample, and the time of each test is recorded. The whole procedure is then repeated, only this time a 5,000-volt negative charge is applied.

For a material to meet the static decay requirements of various military specifications, 100% of the applied charge must be dissipated to ground in 2.0 seconds or less. Because of the difficulties associated with measuring static decay down to zero volts (100% dissipation), a cutoff point at 10% of the applied voltage (90% dissipated) is a frequently used option and represents a reasonable compromise, considering all factors involved. Such things as accurate calibration of the meter, poor reproducibility, and touchy zero adjustments on the equipment are among the factors that make testing down to zero volts difficult.

CONTINUITY TEST

There is also a test to prove whether or not there are any discontinuities in a mat—that is, whether or not grommets are installed tight enough to make good electrical contact with the mat. The proper test for surface resistivity of a table mat is covered in ASTM-D257, and the static decay test in Federal Standard 101, Method 4046. An alternative method using an ohmmeter is also covered in NFPA Standard 56A, available from the National Fire Protection Association (470 Atlantic Avenue, Boston, MA 02210).

Step 1

Place the mat on a nonconductive surface. Use a reliable ohmmeter with sufficient range. A VIZ Senior Voltomyst, Model WV-98C meter with a 1,000-megohm range is acceptable. The meters use one lead with a pointed probe and a second lead with an alligator clip. Fasten a 15.5-ounce weight to the pointed probe. (A standard bar of solder, doubled over and taped to the probe, will do.)

Step 2

Fasten the alligator clip securely to the stud on the mat. Hold the weighted probe by the wire and lower the probe onto the surface of the mat 2 inches away from the stud, allowing the weighted probe to apply the pressure. Record the reading. Repeat this test at various points 2 inches away from the stud in various directions. Record each reading taken. If the mat is equipped with two studs, repeat the entire test around the second stud, recording each reading taken. The average of all readings should be about 200 megohms (1.8 to 2.5×10^8 ohms.)

Step 3

Repeat the same test procedure, but clip the alligator clip to the edge of the mat at random points, and lower the weighted probe anywhere on the face of the mat that is 24 inches away from the point where the alligator clip is located. Remember to use only the weight of the probe to apply pressure to the test point, since additional pressure will cause changes in the values. Take readings from at least ten separate locations on the mat and record them all. The average of all readings should be about 225 megohms (1.9 to 2.7×10^8 ohms).

Step 4

Repeat the test procedure on both diagonals of the mat. The average reading of both diagonals on a 24″ × 36″ mat should be about 250 megohms. If your mat is larger than 24″ × 36″, take two readings on diagonals that are 43.5″ apart. The average reading should be about 250 megohms (2.0 to 3.4×10^8 ohms).

Interpreting Test Results

Should readings taken during Step 2 exceed 350 megohms, in all probability the stud is not properly installed. Ohmmeter readings over 9×10^{11} taken between any two points in Steps 3 and 4 would indicate that the conductive layer in the mat has been damaged, or that the surface of the mat is contaminated. If contamination is suspected, clean the surface using a common dishwashing detergent and repeat the tests. The electrical values given here may vary slightly from manufacturer to manufacturer.

A simple test for evaluating the conductivity of shoes to a grounded floor or conductive mat is shown in Figure 8-9. Any conductive material may be used as the strap around the ankle; a wrist strap is acceptable and is at hand in any static-controlled area. A reading of 1,000 megohms or less indicates that one can step safely onto a grounded floor or mat.

TESTING OPERATORS' CLOTHING

Tests and standards for electrostatic characteristics as a measurable quantity in fabrics are recognized by the National Fire Protection

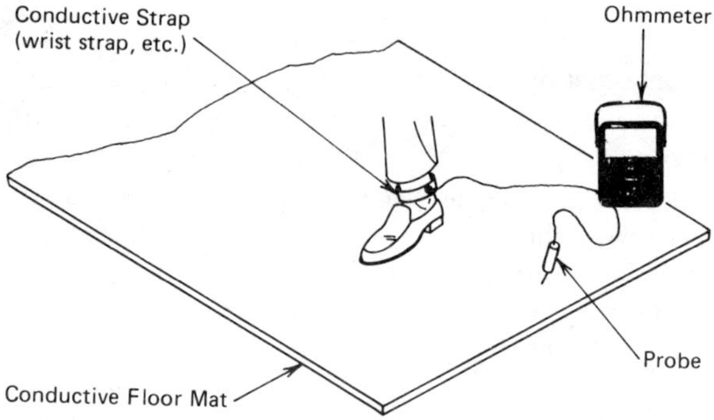

Figure 8-9. Test for conductivity through shoes. The reading should be 1K megohm or less.

Association (NFPA) and set forth in NFPA Standard 56A, applying to areas where inhalation anesthetics are used.

Surface resistivity is tested by an NFPA specified test—NFPA 56A-1973, 4-6.6.3(b)—conducted in accordance with AATCC Test Method 76-1982. NFPA requires that measured resistivity shall be less than 1.0×10^{11} ohms per unit square of material.

Three test specimens are conditioned for four hours at 70° F, 20% relative humidity, and placed in firm contact with concentric rings of 2.75 and 10 inches. 40 volts per centimeter of interelectrode spacing are passed through each sample for a minimum of one minute, or until a constant reading is obtained. The criterion for constant electrical resistivity is less than 0.1 units per minute change in the value of Log R. Results are obtained by averaging the measured resistance values of *N* specimens and computing the resistance on ohms per square as follows:

$$\text{Resistivity in ohms per square} = \frac{2.73 \times \text{measured resistance}}{\log \dfrac{\text{outer electrode radius}}{\text{inner electrode radius}}}$$

The logarithm of the average of the three calculated resistivities is reported in ohms per square of fabric.

Surface resistivity may also be measured by techniques defined by the American Society for Testing and Materials ASTM D257-78. This method has been adopted by the Department of Defense to replace an earlier document: FTM Standard 406, Method 4041. Both documents specify the use of a 500-volt DC test voltage and an electrification time of 60 seconds before recording the resistance value, which is also expressed in ohms per square.

Static charge decay is tested by another NFPA specified test—NFPA 56A-1973, 4-6.6.3(a)—conducted in accordance with Federal Standard 101, Method 4046. This procedure specifies contact charging the material alternately to ± 5,000 volts DC and measuring the decay time to 10% of the maximum charge applied, or 500 volts, in an environment of 70° F ± 3 with a maximum relative humidity of 15%.

For the purpose of ESD-protective materials qualifications, static decay tests should be performed at the lowest relative humidity anticipated in the usage environment. Hygroscopic antistatic materials (those that readily take up and retain water) will show significant variations in decay time with relative humidity changes, particularly at low RH values. Ideally, time to decay to 0 volts would be required. However, because of inaccurate calibration, poor reproducibility, touchy zero adjustment, and so on, time to decay to 10% is most commonly accepted.

The principal documents that govern ESD-control test requirements are listed in Appendix 1 along with a summary of the pertinent requirements. It is wise for the engineer responsible for a static control program to be familiar with the contents of these specifications.

9. MARKING AND IDENTIFICATION

ACCEPTED MARKING SYMBOLS FOR STATIC-SENSITIVE DEVICES

A number of symbols are currently being used throughout the electronics industry as a means of calling attention to static-sensitive electronic equipment, in an attempt to limit the mishandling of these devices. The following figures depict the symbols being recognized as acceptable warning labels and markings.

Figure 9-1 is the official symbol defined by the Electronic Industries Association in their EIA Standard RS-471, and is the recommended symbol. Currently, other symbols are being used by the industry and are being accepted (see Figure 9-2). For applications in military programs, the EIA symbol is recommended. The purpose of this standard is to provide a distinctive symbol and label to be used to identify those solid-state devices that require special handling. The symbol or label should be used at the lowest practical level of packaging and on the device itself, if space permits. It may also be used on device data sheets, on storage bins, and on special protective wrapping materials.

Monochromatic reproduction in any color that contrasts with the background may be used. Where the choice of color is arbitrary, it is suggested that the symbol be black on a yellow background. Wherever possible, the color red for the symbol should be avoided, as red suggests a personal hazard.

The attention label comprises the symbol with the words:

ATTENTION—OBSERVE PRECAUTIONS FOR HANDLING
ELECTROSTATIC SENSITIVE DEVICES

1A – Basic Symbol

1B – Simplified Version For Extreme Reduction

Figure 9-1. Recommended symbols for ESD identification.

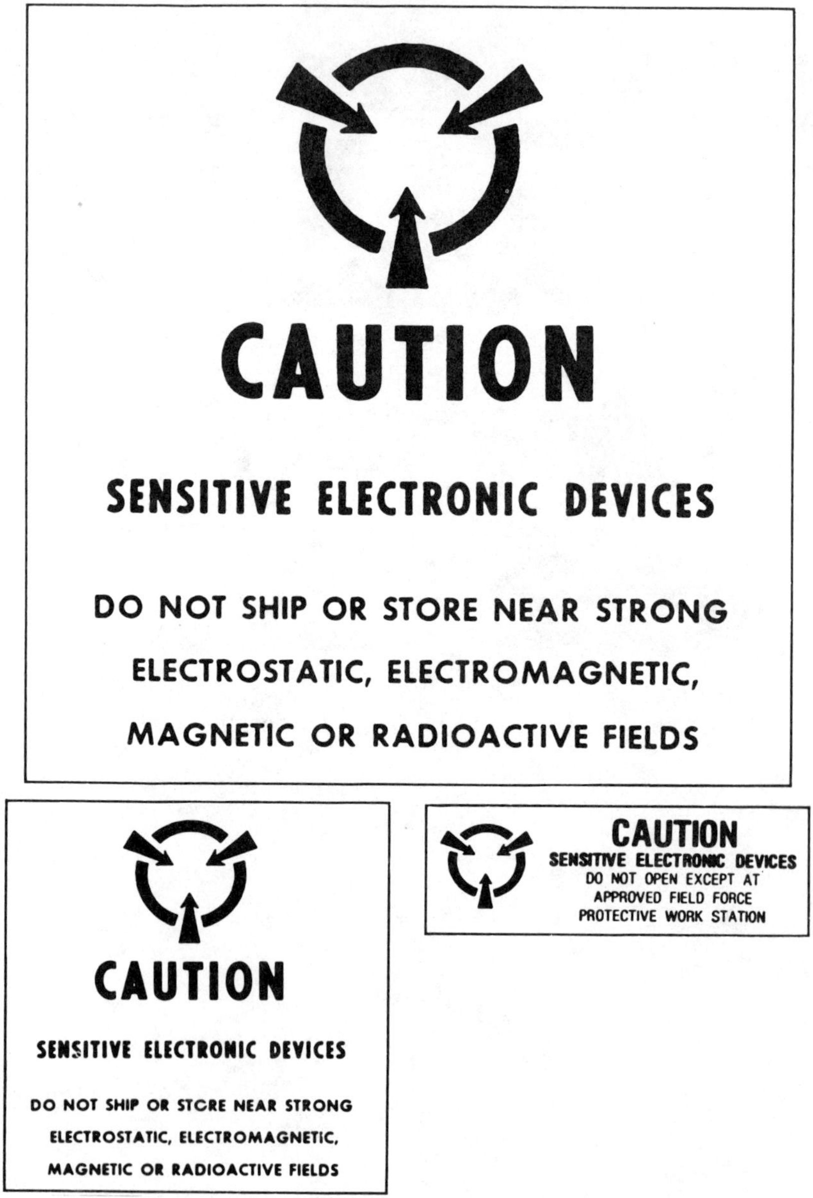

Figure 9-2. Alternate symbol for ESD identification.

APPENDIX 1

Test Requirement and Specification Origins

DOCUMENT	PARAGRAPH	REQUIREMENT
DOD-STD-1686	5.3 and 5.9	Electrostatic voltages in proteted areas must be less than the lowest susceptibility voltage of the parts being handled.
	5.11a	Materials used for ESD protective packaging shall be noncorrosive and nongenerating. Shunts shall be conductive with a maximum surface resistivity of 1×10^5 ohms/square maximum.
	5.11b	ESD-protective packaging must protect contents from electrostatic voltage sources of at least 4,000 V in the form of fields or direct discharge.
DOD-HDBK-263	7.1	Properties of ESD-protective materials: • are nongenerating • provide protection from fields • provide protection from direct discharge
	7.1.1	Resistivity parameter measurements: • volume resistivity (ohm-cm) • surface resistivity (ohms/square) • decay time (sec.)
	7.1.3	Resistivity test methods: • ASTM D-991 • ASTM D-257 • FED-STD 101, method 4046
	7.1.4.1	Triboelectric generation test by rubbing and measuring with a field meter.
MIL-M-38510	5.1 Category A	Antistatic packaging material with surface resistivity not greater than 1×10^{14} ohms/square (no test method specified) and conductive noncorrosive rails shall have surface resistivity less than 1×10^5 ohms/square (no test method specified).
MIL-B-81705	Various	Barrier material, flexible, electrostatic-protective, heat-sealable Type I material is water-vaporproof, greaseproof, electrostatic- and electromagnetic-protective, consisting of metal foil and other laminates meeting electrostatic properties of 2-sec. decay times and EMI attenuation of 25 dB minimum.

Test Requirement and Specification Origins *(cont.)*

DOCUMENT	PARAGRAPH	REQUIREMENT
		Type II material is transparent, waterproof, and electrostatic-protective. It shall be uniform in appearance, tinted any color, and have no talc or other powder substance applied to the surface. It shall meet electrostatic properties of 2-sec. decay.
E.I.A. IS-5	3.2.1.2	Magazines: antistatic Faraday-cup test: maximum based on part and package type leaded MOS: 30 pc./ld; leaded bipolar: 50 pc/ld; unleaded MOS: 10 pc/ld; unleaded bipolar: 20 pc/ld.
	3.2.2.2	Magazines: shielding • End-to-end resistance test: 1×10^6 ohms/square • Surface resistivity test: 1×10^5 ohms/square (ASTM D-257) • Volume resistivity test: 1×10^4 ohm-cm (ASTM D-991)
	3.3	Bags and boxes Antistatic type: • Decay test: 2-sec. max. (101/4046) • Surface resistivity: $1 \times 10^5 < p_s < 1 \times 10^{13}$ (ASTM D-257) Conductive type: • Surface resistivity: 1×10^5 ohms/square (ASTM D-257) • Volume resistivity: 1×10^4 ohm-cm (ASTM D-991) Shield type (continuous and inside antistatic): • Surface resistivity: 1×10^5 ohms/square (ASTM D-257) • Volume resistivity: 10 ohm-cm (D-991) • Inaccessible shield-plate capacitor sensor test: Class I to IV: 100 V to 2,000 V
	3.4	Flexible and rigid cushioning material Antistatic type: • Decay test: 2 sec. max. to 1% (101/4046) Conductive type: • Volume resistivity: 1.0×10^4 ohm-cm (ASTM D-991)
	3.6	Loose-fill, molded, or irregularly shaped materials: Antistatic type: • Horsehair/Faraday cup: 50 pc/square in max.

Standard Test Methods

TEST METHOD AND TITLE	PROPERTIES MEASURED	INTENDED APPLICATIONS
FED-STD 101, test method 4046.1, electrostatic properties of materials	Decay time	Materials in film or sheet form
ASTM D257-78, standard test method for DC resistance or conductance of insulating materials	DC insulation resistance, volume resistance, volume resistivity, surface resistance surface resistivity	Electrical insulating materials
ASTM D2679-73, standard test method for electrostatic charge	Surface resistivity transferred charge per unit area (or per unit volume or mass) charge decay rate	Charge transferred between two objects in contact; useful for solid sheet, film, fabric, powder, liquids, and gases
ASTM D3509-76, standard test method for electrostatic field strength due to surface charges	Electrostatic field strength	Metal objects of insulating bodies with electrostatic charge
ASTM D991-82, standard test method for rubber property-volume resistivity of electrically conductive and antistatic products	Volume resistivity (assuming surface conductivity is negligible in comparison)	Rubbers used in electrically conductive and insulative products
E.I.A. IS-5, para. 3.8, shunting foam	Pin-to-pin resistance	Conductive shunting foam
E.I.A. IS-5, Appendix A1, Faraday-cup test	Electrostatic charge	Magazines (slides, tubes, rails)
E.I.A. IS-5, Appendix A2, test for electrostatic shielding on inaccessible shielding surfaces	Peak voltage across capacitor sensor	Electrostatic shielding bags, pouches, and sheets (non-cushioning)
E.I.A. IS-5, Appendix A3, triboelectric testing for irregularly shaped items or materials	Electrostatic charge	Antistatic type: loose-fill, molded, or irregularly shaped packaging materials
AATCC test method 134-1975, electrostatic propensity of carpets	Peak voltage from step test; peak voltage from scuff test	Carpets
NFPA standard 56A, para. 4-6.2.7b, inhalation anesthetics	Surface resistance	Conductive flooring

ESD Protective Packaging: Material Classifications

CLASSIFICATION	STANDARD PARA. NO.	DEFINITION
Antistatic	DOD-HDBK-263 Para. 3.1	Defines antistatic material having surface resistivity $> 1 \times 10^9$ ohms/square and $< 1 \times 10^{14}$ ohms/square.
	MIL-M-38510 Para. 3.1.3.2.1	Defines "antistatic" material having surface resistivity $< 1 \times 10^{14}$ ohms/square and resisting charge generation.
	E.I.A.-IS-5 Para. 2.2	Defines "antistatic" materials as those that resist charging from contact separation and [classified] into two types: (1) impregnated types and (2) surface-treated types.
	Para. 3.3.1.2(2)	Surface resistivity: $1 \times 10^5 < p_s < 1 \times 10^{13}$ ohms/square.
Static-dissipative	DOD-HDBK-263 Para. 3.17	Defines "static-dissipative" material having surface resistivity $> 1 \times 10^5$ ohms/square and $1 < \times 10^9$ ohms/square.
Conductive	DOD-HDBK-263 Para. 3.2	Defines "conductive" material having a surface resistivity of 1×10^5 ohms/square maximum.
	MIL-M-38510	Defines "conductive" material having a surface resistivity 1×10^5 square/ohms or volume resistivity of 10^3 ohm-cm maximum.
	E.I.A.-IS-5	Defines "conductive" materials as those that are either metal or impregnated with metal or carbon or other conductive material, or whose surface has been treated with such. Also states charges can be produced by these materials when separated from nonconductive surfaces.
Electrostatic-shielding	E.I.A.-IS-5 Para. 2.4	Defines "electrostatic-shielding" materials as those capable of attenuating an electrostatic field so that its effects do not reach the stored item.

APPENDIX 2

GLOSSARY

Alpha particle. A positively charged particle consisting of two protons and two neutrons emitted from the nucleus during a decay process. Because of its size, it has poor penetrating ability. It cannot penetrate skin, but can ionize air molecules.

Ampere. Practical unit of electrical current. One ampere of current is 6.24×10^{18} electrons passing one point in one second. Abbreviated as amp. 1 amp = 1 coulomb/second.

Bag. A preformed container made of flexible material generally enclosed on all sides except one which forms an opening that may or may not be sealed after loading. It is normally constructed from one piece of material that has been folded over and sealed on two edges.

Bipolar device. A semiconductor device which is current- or power-sensitive and can be damaged by coming into contact with a charged body causing a transient discharge to pass through the junction of the device. This discharge produces such a high current density at the junction that overheating and melting occur.

Capacitor/capacitance. The ability of a component or material to store an electric charge. the capacitance of a charged conductor is the ratio of its charge to its voltage (i.e., $C = Q/V$). Capacitance is measured in terms of *farads.* Since the farad is such a very large number, its values are usually expressed in millionths of a farad or *microfarads,* or millionths of millionths of a farad or *picofarads.*

Carrier. Holder for electronic parts and devices which facilitates handling during processing, production, imprinting, or testing operations and protects such parts under transport.

Charge. Measured in coulombs or fractions thereof. The static charge on a body is measured by the number of separated electrons on the body (negative charge), or the number of separated electrons not on the body (positive charge). Since electrons cannot be destroyed, an electron removed from one body must go to another body, leaving behind a positive (+) void. Thus equal and opposite charges are always produced.

Conductor (conductive). A substance or body that allows a current of electrons to pass continuously along it or through it when a voltage is applied across any two points. Such materials exhibit relatively low resistance.

Coulomb. A specific quantity of electrons (charge) on a body, expressed as 1 coulomb = 6.24×10^{18} electrons. In electrostatics a much more practical unit is the *nanocoulomb* (10^{9} coulombs), representing a charge of 6.24×10^{9} electrons.

Current. The flow of electrons past a certain point in a specified period of time, measured in amperes or fractions thereof. Current is measured in terms of electrons per second, but since this number would be tremendously large, it is usually stated in terms of coulombs per second. 1 coulomb per second = 1 ampere.

Dielectric breakdown. A threshold effect in a dielectric medium where, at some electric field strength across the medium, bound electrons become unbound and travel through the medium as a current. In solid media, the region of the current is permanently damaged. The units of measurement are usually volts per unit of thickness.

DIP. Dual-in-line-package. A type of housing for integrated circuits, the standard form of which is molded plastic with two rows of pins. These packages are also made of ceramic with metal or ceramic ends or lids. It is a more popular type of containment than the flat pack or TO-Can for industrial use, because it is relatively inexpensive and easily wave-soldered into printed circuit boards.

Electrically continuous. A surface that is electrically conductive, in that current can be passed as the result of an applied voltage between any two points on its physical surface, and discontinuities, slots, or holes do not occupy more than 10% of the material's surface.

Electrically powered. Electrically powered static-elimination equipment is made up of two basic components: the static eliminator, which generally consists of one or more electrified needles rigidly held less than an inch from a grounded metal housing or proximity ground; and the high-voltage supply which powers the static eliminator. Ion generation from electrical static eliminators occurs in the air space surrounding the highly charged needle points.

Electromagnetic shield. A screen or other housing placed around devices or circuits to reduce the effect of both electric and magnetic fields on or by them. The electromagnetic field results from the presence of a rapidly moving electric field (rf) and its associated magnetic field. Shielding from electromagnetic interference (abbreviated as EMI) is a combination of reflection and absorption of electromagnetic energy by the material. Reflection occurs at the surface much like the reflection of light at an air-to-water interface, and is not usually affected by shield thickness. Absorption, however, occurs within the shield and is highly dependent on thickness. The best electromagnetic shielding materials are ferritic, i.e., steel, nickel, etc. Aluminum is often used in less critical situations, however.

Electron. A negatively charged particle with an electrical charge equal to about 1.6×10^{19} coulomb.

Electrostatic field. The region surrounding an electrically charged object in which another electrical charge can be induced and will experience a force. Quantitatively, it is the voltage gradient between two points at different potentials.

Electrostatic shield. A barrier or enclosure that prevents the penetration of an electrostatic field. An electrostatic shield, however, may not offer much protection against the effects of electromagnetic interference (EMI). EMI shields, however, are good electrostatic shields.

EMI. Electromagnetic interference. Sources are static sparks, lightning, radar, radio and TV transmission, brush motors, line transients, etc. By line conduction or air propagation, EMI can induce undesirable voltage signals in electronic equipment causing malfunction and occasionally component damage. Protection against EMI usually requires the use of shields, filters, and special circuit design. See also **Electromagnetic shield, Electrostatic shield,** and **Radio frequency interference.**

Energy. Measured in *joules* or fractions thereof. A spark is energy being expended. The measure of energy takes several forms. If it is electrical energy, it is measured in watt-seconds or joules. Since a joule is quite a bit of energy, and static sparks do not usually have this much energy, their energy is usually measured in thousandths of a joule (*millijoules*). A static spark in the order of millionths of a joule (*microjoules*) will damage semiconductors. Electrostatic energy is calculated from the relations $CV^2/2$, $Q^2/2C$ or $QV/2$.

Faraday cage. An electrically continuous, conductive enclosure which provides electrostatic shielding. The cage or shield is usually grounded, although it need not be.

Ground. A metallic connection with the earth to establish zero potential or voltage with respect to ground or earth. It is the voltage reference point in a circuit. There may or may not be an actual connection to earth, but it is understood that a point in the circuit said to be at ground potential could be connected to earth without disturbing the operation of the circuit in any way. Grounds that can be used for static control work stations include water pipes, any power ground, or any large metal structural member of a building.

Grounding. Connecting to ground or to a conductor that is grounded. A means of referencing all conductive objects to a zero voltage equipotential surface. This is the surest method of eliminating ESD, since everything is maintained at the same potential.

Induction static eliminator. Generally, a series of conductive grounded points or brushes. When a single sharp grounded needle point is brought into the proximity of any highly charged surface, it has induced in it a charge opposite to that of the surface. When a high enough charge concentration has been developed, the surrounding air will break down. A vast number of charge-balancing ions are formed. The simple "tinsel" static eliminator is an example of an induction static eliminator.

Insulator. A material that does not conduct electricity. A nonconductor.

Integrated circuit. Abbreviated IC. Any electronic device which contains transistors, resistors, capacitors, etc., in a single package. To further define the term, additional modifiers are often prefixed. For instance, a bipolar IC is one containing junction semiconductors that use both positive (p-type) and negative (n-type) charge carriers.

Ionization. The process by which a neutral atom or molecule, such as air, acquires a positive or negative charge.

Joule. A unit of energy. The energy of a static discharge is $\frac{1}{2} CV^2$ where C = capacitance of the discharging object, and V = voltage difference between the discharging object and the point to which it discharged.

MOS device. A device having a metal-oxide semiconductor layer structure which is voltage- or electric-field-sensitive and can be damaged by a static discharge or by induction from an electrostatic field. The damage occurs as a dielectric breakdown when the voltage or electric field across the dielectric layer between two relatively conductive layers exceeds the dielectric strength of that material.

Nonconductor. See **Insulator.**

Nuclear static eliminator. Nuclear static eliminators create ions by the irradiation of the air molecules. Some models use a safe alpha-emitting isotope to create sufficient ion pairs to neutralize a charged surface. The high-speed particle interacts with air molecules with sufficient energy to actually strip off one of its outer electrons (see **Ionization**).

Ohm. The unit of electrical resistance. It is the resistance through which a current of one ampere will flow when a voltage of one volt is applied.

Package. A package is the enclosure of products, devices, or other packages in a wrap, pouch, bag, slide, magazine, or other container form so as to perform one or more of the following functions:

1. Containment for handling, transportation, and use.
2. Preservation and protection of the contents for the life of the item.
3. Identification of contents, including quantity and manufacturer.
4. Facilitate the dispensing and use of the contents.

Packaging. In the electronics industry, packaging refers to the process of locating, connecting, and protecting various devices, components, etc. For example, entire circuits can be printed onto thin film wafers which are electrically interconnected during fabrication. These miniature circuits are then fused into an easy-to-handle element called a package. This package both protects the semiconductor circuitry and permits convenient external connections to be made to it.

Potential. Measured in millivolts, volts, or kilovolts. Potential or voltage is measured from a base point. This point can be any voltage but is usually ground, which is theoretically zero voltage.

Pouch. A small or moderately sized baglike container constructed by the sealing on three edges of two flat sheets of flexible material, or by sealing one end of a tube of flexible material.

Radio frequency interference. Abbreviated as RFI. A form of electromagnetic interference (EMI). Any electrical signal capable of being propagated and of interfering with the proper operation of electrical or electronic equipment. The frequency range usually includes the entire electromagnetic spectrum. The spark from a static discharge is a source of RFI.

Resistance. Electrical current encounters difficulty in passing through an electrical circuit or conductor. This difficulty can be measured and is called *resistance.* This is a bulk property of a material and depends on the material's dimensions, electrical resistivity, temperature, and also voltage in non-ohmic materials. The resistance of the material determines the current (electron flow) produced by a given voltage. The practical unit of resistance is the *ohm.*

Resistivity. A measure of the intrinsic ability of a material to conduct current. Its value is independent of the dimensions of the material. Both conductors and nonconductors have resistivity. The unit of volume resistivity is the *ohm-cm.* The unit of surface resistivity is *ohms per square.*

Sensitivity. The minimum value which a sensor will effectively and reliably detect.

Surface resistivity. The ratio of DC voltage to the current that passes across the surface of the system. In this case, the surface consists of the geometric surface and the material immediately in contact with it. In effect, the surface resistivity is the resistance between two opposite sides of a square and is independent of the size of the square and its dimensional units. Surface resistivity is relevant only for materials where current conduction is virtually only on the surface. Surface resistivity is not meaningful for volume-conductive materials. The unit of surface resistivity is *ohms per square.* The unit *ohms* is occasionally used, since the value is the resistance across a square. The use of ohms per square is recommended to distinguish surface resistivity from arbitrary resistance.

Triboelectric. Pertaining to an electrical charge generated by frictional rubbing or separation of two surfaces.

Triboelectric series. A list of substances arranged so that any of them can become positively charged when rubbed with one further down the list, or negatively charged when rubbed with one further up the list. Generally, the further apart such materials are in the triboelectric series, the greater their tendency to charge one another. This series is derived from specially prepared and cleaned materials tested in very controlled conditions. In everyday circumstances, materials reasonably close to one another in the series can produce charge polarities opposite to that expected. This series is only a guide. See Figure 1.

Volt. The unit of voltage, potential, and electromotive force. One volt will send a current of one ampere through a resistance of one ohm. Abbreviated as V.

Voltage. Term used to designate the electrical potential difference that exists between two points and that is capable of producing a flow of current when a closed circuit is connected between the two points.

Voltage suppression. A phenomenon where the voltage from a charged object is reduced by increasing the capacitance of the object rather than decreasing the charge on the object. The relation $Q = CV$ describes the phenomenon. It occurs most frequently when a charged object is close to a ground plane, but not in resistive contact with the ground plane.

Volume resistivity. The ratio of the DC voltage per unit of thickness, applied across two electrodes in contact with or embedded in a specimen, to the amount of current per unit area passing through the system. Volume resistivity is generally given in *ohm-centimeters.*

APPENDIX 3

SOURCES OF ADDITIONAL INFORMATION, PARTS, AND MATERIALS

1. Unger, B., R. Chemelli, P. Bossard, M. Hudock, "Evaluation of Integrated Circuit Shipping Tubes." *EOS/ESD Symposium Proceedings,* 1981.
2. Bossard, P. R., R. G. Chemelli, B. A. Unger. "ESD Damage from Triboelectrically Charged Pins." *EOS/ESD Symposium Proceedings,* 1980.
3. Unger, B. A., "Electrostatic Discharge Failures of Semiconductor Devices." *IEEE Reliability Physics Symposium Proceedings,* 1981.
4. Huntsman, J. R., D. Yenni, Jr., G. Mueller. "Fundamental Requirements for Static Protective Containers." 1980 Nepcon/West Conference.
5. Speakman, T. S., "A Model for Failure of Bipolar Silicon Integrated Circuits Subjected to Electrostatic Discharge." *IEEE Reliability Physics Symposium Proceedings,* 1974.
6. Halprin, S. A., "Static Control Using Topical Antistats." *EOS/ESD Symposium Proceedings,* 1979.
7. Scovern, J. L., "Reducing Solder Defects Using Topical Antistats." *Circuit Manufacturers Magazine,* April 1981. International Printed Circuits Conference, West, 1980.

ADDITIONAL HELP

1. Fuqua, Norman B. "ESD Protective Material and Equipment: A Critical Review." Reliability Analysis Center state-of-the-art report, Spring 1982, p. 16.
2. Antonevitch, John N., and Mark Blitshteyn. "Measuring Effectiveness of Air Ionizers." Reliability Analysis Center, EOS/ESD Symposium, 1983, p. 77.
3. Mykkanen, C. Fred, and David R. Blinde. "The Room Ionization System: An Alternative to 40 Percent RH." *Evaluation Engineering,* September 1983, p. 86.
4. Donabach, Kurt W., and Morton Walter. "What You Always Wanted to Know About Negative Ions." The International Institute of Natural Health Sciences, Inc.

Recommended Reading

1. Yenni, D. M., Jr., and J. R. Huntsman. "The Deficiencies in Military Specification MIL-B-81705: "Considerations and a Simple Model for Static Protection." Reliability Analysis Center, EOS/ESD Symposium, 1979.

2. Military Specification MIL-M-38510, Revision E, Section 5.
3. Yenni, Don M., Jr. "Basic Electrical Considerations in the Design of a Static-Safe Work Environment." 1979 Nepcon/West Conference Anaheim, Calif., p. 2.
4. Gruhalla, Michael E. "Electrical Grounding in Large Instrumentation Systems." Reliability Analysis Center, EOS Symposium, 1982.

Specification References

1. DOD-HDBK-263, May 2, 1980
2. DOD-STD-209
3. FED-STD-209
4. MIL-B-81705
5. MIL-M-38510, Revision E
6. MIL-M-55565
7. MIL-STD-883B
8. Defense Standard 59-98, Ministry of Defense, Crown, 1979.

Manufacturers' Index

The following manufacturers' Index is a starting point for users shopping for ESD-protective devices:

Charleswater Products, Inc.
93 Border St.
West Newton, MA 02165

Products:
- Statguard floor finish
- Micastat amino resin tabletop laminate
- Statshield conductive transparent bags
- Statfree conductive foam

Semtronics Corp.
P.O. Box 592
Martinsville, NJ 08876

Product:
- Enstat ribbed conductive rubber and vinyl floor mats

Vinyl Plastic Inc.
P.O. Box 451
Sheboygan, WI 53081

Product:
- Conductile conductive vinyl tile

Simco Co., Inc.
2257 N. Penn Rd.
Hatfield, PA 19440

Product:
- Electrical source ionizers

*Static Control Systems/3M
22-2SW, 3M Center
St. Paul, MN 55144

Products:
- 8200 series table mats
- 2068 series wristbands
- 2100 series conductive transparent bags
- Velostat conductive bags
- Nuclear source ionizers
- Conductive tote boxes
- Conductive foam

*Olympic Plastics Co., Inc.
5800 W. Jefferson Blvd.
Los Angeles, CA 90018

Product:
- Protect-O-Stat conductive tote box

Meritex Plastic Industries, Inc.
3301 E. Randol Mill Rd.
Arlington, TX 76011

Product:
- Antistatic DIP and TO-3 tubes

ADE Inc.
1560 E. 98th St.
Chicago, IL 60628

Product:
- Cancel-3 antistatic cushioning material

Angelica Uniform Group
700 Rosedale Ave.
St. Louis, MO 63112

Products:
- Freostat smocks and lab coats
- ASQ-100 clean-room garments

Fluorware
Jonathan Industrial Center
Chaska, MN 55318

Products:
- Stat-Pro 100 wafer carriers
- Stat-Pro 150 chip trays

Supplier Guide

*Bemis Co., Inc.
Minneapolis, MN

*Bengal, Inc.
Sepulveda, CA

Buckhorn Corp.
Material Handling Group
Cincinnati, OH

C.R. Daniels, Inc.
Ellicott City, MD

Charleswater Products, Inc.
West Newton, MA

Conductive Containers, Inc.
Deefield, IL

Contact East
Burlington, MA

Control Static Div., ISCS
Santa Fe, CA

Gemini Plastic Enterprises, Inc.
Maywood, CA

Glen-Mitch Tool, Inc.
Schaumber, IL

ISC Santa Clara, CA
Santa Clara, CA

Julie Associates, Inc.
Billerica, MA

Lewisystems/Menasha
Watertown, WI

Maine Poly, Inc. Greene, ME
Green, ME

Max Carrier Systems
Multi Tool and Mfg.
Cape Canaveral, FL

Olympic Plastics Co., Inc.
Los Angeles, CA

Panel Controls Corp.
Detroit, MI

Protecta-Pack Systems
Minneapolis, MN

Richmond Div., Dixico, Inc.
Redlands, CA

Semtronics Corp.
Scotch Plains, NJ

The Simco Co., Inc.
Hatfield, PA

*3M, Static Controls
Systems Division
St. Paul, MN

*Provided assistance in the preparation of this text.

The following companies and organizations offer products or services in the ESD field:

ACL
Charleswater Products
3M
Instrument Specialties
Plastic Systems
Reliability Analysis Center
SAR
Sealed-Air Corp.
Simco
Static, Inc.
Alacra Systems
American Convertors
Bemis Company
Conductive Containers
Fancort Industries
L. Gordon Packaging
Stephen Halperin & Assoc.
Herman Miller, Inc.
Jiffy Packaging
Lewisystems
Maine Poly
3M, Static Control Systems Div.
Seco Industries
Sentinel Foam Products
Simco

INDEX